ISBN 978-1-330-06673-7
PIBN 10017302

1 MONTH OF FREE READING

at

www.ForgottenBooks.com

English
Français
Deutsche
Italiano
Español
Português

www.forgottenbooks.com

Mythology Photography **Fiction**
Fishing Christianity **Art** Cooking
Essays Buddhism Freemasonry
Medicine **Biology** Music **Ancient**
Egypt Evolution Carpentry Physics
Dance Geology **Mathematics** Fitness
Shakespeare **Folklore** Yoga Marketing
Confidence Immortality Biographies
Poetry **Psychology** Witchcraft
Electronics Chemistry History **Law**
Accounting **Philosophy** Anthropology
Alchemy Drama Quantum Mechanics
Atheism Sexual Health **Ancient History**
Entrepreneurship Languages Sport
Paleontology Needlework Islam
Metaphysics Investment Archaeology
Parenting Statistics Criminology
Motivational

COMMERCIAL
GEOGRAPHY

BY

ALBERT PERRY BRIGHAM

GINN AND COMPANY
BOSTON · NEW YORK · CHICAGO · LONDON

The Athenæum Press

GINN AND COMPANY · PRO-
PRIETORS · BOSTON · U.S.A.

PREFACE

The object of this text is the exposition of the principles of commercial geography as based on a knowledge of its more important facts. It is regarded as necessary and advantageous to treat products and regions in a single volume, if perspective is observed and if both products and regions are made contributory to the unfolding of principles. To this end minor commodities and regions must surrender their place to facts of larger meaning.

Part I offers an inductive approach to principles which are formally stated in Chapter VI. The author holds it wise to avoid an introductory statement of abstract relations, and has therefore chosen five products or staples of world-wide interest, treating them broadly as world products and as typical of all others in the geographic principles involved. No materials of commerce are more significant in themselves, or in their relations, than wheat, cotton, cattle, iron, and coal.

Part II relates to the United States and opens with a brief review of physical features. To the usual account of plant, animal, and mineral substances, a chapter on water is added. This is amply justified by the importance which water has now assumed as a part of our natural resources. The chapters on concentration of industry, centers of general industry, transportation, communication, and government relations afford a return to vital principles, unfolding them more fully than was possible in Chapter VI, and offering considerations which are of universal application, though here developed in special reference to our own country.

Part III deals with foreign countries. Canada is taken first, owing to its close geographic and commercial relations to the United States. It is followed by the chapters on the great industrial and trading nations of western Europe. Grouping the

iii

minor with the greater countries, southern and eastern Europe follow, and a single chapter is given to each of the remaining continents. For an elementary text to be used in American schools it is believed that this allotment of space is wise. The closing chapter summarizes the history of commerce and suggests some of its larger aspects.

It has been sought to place orderly and cumulative emphasis on general principles; to concentrate, so far as possible, the treatment of each topic; to use sparingly statistics of temporary value; to give little attention to industrial processes except as they have geographic meaning; and to present industry and commerce as organic, evolutionary, and world-embracing, responding to natural conditions and to the spirit of discovery and invention, and closely interwoven with the higher life of man.

The illustrations consist of views, diagrams, and maps. These stand in close relation to the text, and the maps are so planned that each may exhibit one or a few things in a legible and simple manner.

The thanks of the author are due to Mr. Chester M. Grover of the High School of Commerce, Boston, who has read the proofs and made many welcome suggestions; and to Mr. R. J. H. Deloach, Professor of Cotton Industry in the Georgia State College of Agriculture, who has performed a similar service for the chapter dealing with cotton. Obligation is acknowledged also to Professor G. G. Chisholm of the University of Edinburgh, whose "Handbook of Commercial Geography" and whose numerous special papers are useful to every worker in this field.

Among those who have aided in the illustration of the volume, thanks are given to The University of Chicago Press for permission to use several maps from the series of base maps prepared by Professor J. Paul Goode; to Professor J. McFarlane of the Victoria University of Manchester; Mr. George L. Buck, Chicago; Mr. James Warbasse, Gloversville, New York; Dr. Charles F. McClumpha, Amsterdam, New York; Dr. L. A. Bauer, Director of the Department of Terrestrial Magnetism,

PREFACE

Carnegie Institution; and Professor Ralph S. Tarr of Cornell University. To Messrs. Gregory, Keller, and Bishop of Yale University the author is indebted for permission to use, in slightly modified form, several maps from their " Physical and Commercial Geography " ; these are, Physiographic Regions of the United States, New York Harbor and its Approaches, and the maps showing the ocean routes of the world and the routes and centers of ancient commerce. Many government bureaus and the officers of several agricultural experiment stations have rendered generous assistance.

ALBERT PERRY BRIGHAM

CONTENTS

PART I

INTRODUCTION TO COMMERCIAL GEOGRAPHY

PART II

COMMERCIAL GEOGRAPHY OF THE UNITED STATES

PART III

COMMERCIAL GEOGRAPHY OF OTHER COUNTRIES

ILLUSTRATIONS, DIAGRAMS, AND MAPS
IN BLACK AND WHITE

LIST OF COLORED MAPS

COMMERCIAL GEOGRAPHY

PART I. INTRODUCTION

CHAPTER I

WHEAT

If one should follow a handful of wheat from the yellow field, by wagon and freight car or ship, to the flouring mill, the provision merchant, the bake oven, and the loaf of bread, he would understand one of the chief themes and acquire many of the principles of commercial geography. Food is the first need of man, and wheat, which has been called the "international grain," is perhaps the most important of foods. We therefore study wheat for itself and for its general illustration of the laws of production and exchange.

1. History of wheat. Wheat, like other cereals, belongs to the order of grasses. It has been modified from some wild grass, but the time when this improvement took place is beyond the memory of man, and the wheat plant as we know it has never been found growing in a wild state. Some scholars think it had its beginning in western Asia and spread eastward to China and westward to Egypt and the countries of Europe. The Swiss lake dwellers raised wheat before the days of written history, and it grew in the valley of the Nile from ancient times to the classic days when "corn" ships sailed from Alexandria to Rome. Wheat growing has now spread over Europe wherever the climate permits, and the grain was brought in the sixteenth century to North America. As new lands have been subdued by civilized man, wheat has moved into the temperate regions of the southern hemisphere and is now an important crop in South America, South Africa, and Australia. The wheat map of the world (fig. 12) shows two irregular belts of wheat countries in the temperate latitudes on both sides of the equator.

2. Varieties of wheat. In the earliest times all wheats may have been alike, but differences of soil, climate, and culture tend to make kinds of wheat with distinct qualities, each breeding true for a long period. Thus there are red and white, bearded. and bald, winter and spring wheats. After wheat became very

FIG. 1. Plowing for wheat in Manitoba

important, men began to breed varieties by crossing good sorts in the hope of producing something better. Wheats were better if they were suited to a wider range of climate, if they gave larger yield, could resist disease and pests, or had higher value for food.

In Germany new varieties of beet have increased the amount of sugar that can be obtained from a field. In California, Luther Burbank has bred remarkable fruits and flowers. In the Minnesota Agricultural Experiment Station and other similar stations new wheats are raised and samples of seed sent out to farmers for testing. The Canadian government has a farm at Ottawa, on which special effort is made to breed wheats that will mature in a short season, in order to extend the Canadian wheat belt as far north as possible. The Department of Agriculture at Washington sends "agricultural explorers" to all lands to seek new and useful plants. They send home wheats that thrive and yield well in lands of small rainfall in the Cordilleran region.

3. Climates and soils suited to wheat. This grain does not thrive in very hot or very cold regions. It needs a cool and moist period for germination and early growth, but it matures best in bright and comparatively dry weather. We can thus

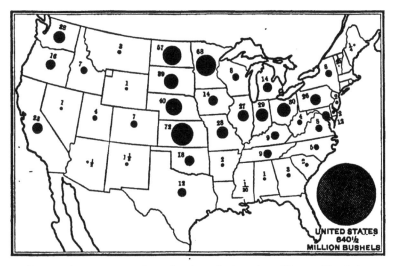

FIG. 2. Average annual production of wheat in millions of bushels,
1899–1908

understand why Egyptian and American wheats are bright, plump, and valuable. A cover of snow is favorable to a good crop of winter wheat, while an open winter, with exposure of the roots to severe changes, is harmful. The soil should be neither too light (sandy) nor too heavy (clayey), but loamy and well drained, with a surface suited to modern plowing and reaping machines. Such conditions are best met on the great plains of temperate latitudes, as on the prairies of the United States, the plains of Canada, the pampas of Argentina, and the plains of southern Russia. Wheat requires warmer summers than rye, barley, or oats, and hence is not found as far from the equator as these grains. It does not need as much heat as Indian corn,

FIG. 3. Spring wheat grown at the United States Experiment Station, Sitka, Alaska, latitude 57 degrees

and thus the center of the corn belt is further to the south than that of wheat.

In North America wheat reaches from southern Mexico to Alaska and to Fort Simpson on the Mackenzie River (62 degrees), but the bulk of the crop lies between 35 degrees (Chattanooga, Memphis, Little Rock) and 55 degrees (central

Saskatchewan and Alberta). In the Old World wheat extends to Trondhjem, Norway, latitude 63 degrees, and in Russia to 60 or 62 degrees. On the south it crosses the Mediterranean to Algiers, about 36 degrees. In India wheat grows south of the tropic of Cancer to about 21 degrees. The high slopes and plateaus of India carry wheat in that country almost as far south as the city of Mexico, and give this cereal a latitude range of over 40 degrees in Eurasia.

The traveler need not fail to find a ripened field of wheat in any month of the year. He may start in November in South Africa, proceed in December and January to Australia, cross the sea to India in February and March, and go on to Egypt in March or April. In May he can visit Algiers, in June Italy and Spain, in July and August central Europe, and in September and October he will discover the harvest in Scotland, Scandinavia, and Finland.

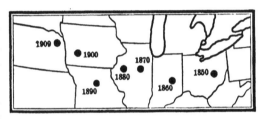

FIG. 4. Movement of the center of wheat production in the United States, 1850–1909

Even in North America the harvest lasts from April, in Mexico, to August in eastern Canada.

4. Progress of wheat culture in North America. In early days wheat was raised only on the Atlantic seaboard, to supply the scattered colonists. After the Revolution, settlers pressed into western New York, and upon the completion of the Erie Canal the " Genesee country " became the wheat center of the United States. As the population grew to the westward, wheat fields covered more and more of Ohio, Indiana, and Michigan, and extended over the prairies of Illinois and southern Wisconsin. In 1880 wheat growing had passed across the Mississippi River as far as central Kansas and Nebraska, and occupied a narrow strip of Dakota along the Red River of the North. In 1900 the center of the wheat areas of the United States was in Iowa, seventy miles west of Des Moines, and North Dakota had become

one of the greatest producers of wheat, ranking with Minnesota
and Kansas as one of the three greatest in the Union. For
many years also large crops have been raised in the three states
of the Pacific coast.

Meantime the older states, like Pennsylvania and Ohio, still
raise much wheat, and the center of the wheat lands moves to
the northwest only because such vast crops are raised in that
section of the country. As population grows, more land is required for pasture, forage, and garden crops, and the largest
wheat-growing seeks new lands. For this reason many American farmers are passing into Canada, and there, with other immigrants, on the great plains of Manitoba, Saskatchewan, and

FIG. 5. Wheat harvest in Manitoba

Alberta, they may develop the greatest wheat center of North
America. The wheat crop of the United States in 1906 was
735,000,000 bushels. Forty years before, in 1866, the crop
was 232,000,000 bushels. This increase was needed partly to
feed our own greater population and partly for export to other nations, which, like England, can raise but a part of their own bread.

5. Wheat on the farm. In the earliest days the methods of
the farmer were rude and simple. He dug the ground with a
pointed stick or a mattock, sowed the wheat " broadcast " out of
a sack, and covered it by means of a harrow. He harvested it
with a sickle and threshed it by the hoofs of beasts or with a
flail. These primitive ways had not been greatly improved by

the middle of the last century, except by the introduction of the plow for tillage and the cradle for harvesting. Then threshing machines, run by the treading of a pair of horses, gave way to machines of 10 horse power, turning out 500 bushels in a day. Reaping machines came in, and four or five men followed each bout to bind the sheaves. Then followed "self-binders," and drills for sowing, by which a single team could sow several acres in a day, covering the seed to even depths in mellow soil.

As wheat spread across the prairies to the Great Plains and the valley of California, whole plantations were devoted to wheat, —

FIG. 6. Wheat in the shock on the prairies

"bonanza" farms of thousands of acres. Broad plains, free from stone, made yet larger machinery possible. Gang plows make swift work of turning the soil. The harrow that fits the soil for the seed may be 25 feet wide. The drill, drawn by four horses, is 12 feet wide. The harvester may cut a swath of 20, 30, or even 50 feet. In some cases the harvester is a thresher as well. In front of the machine the wheat is standing; behind, it lies on the ground in sacks ready for market. "The enormous size of the machine; the team itself, — twenty large mules driven in two ranks, ten abreast; the separator towering high above the standing grain; the loud hum of the monster as it sweeps over the plain; the cloud of dust floating away on the breeze; the broad swath and the sacks of grain gliding from its side and strewed along the way, made it, all in all, the most impressive piece of machinery I have ever seen at work." [1]

Western wheat is not all raised and harvested on a giant scale, for there are many farms, even in such a state as Kansas, of not more than three hundred acres, on which the owner, with a few helpers and smaller machines, raises his crop. On great frontier farms wheat is raised every year on the same fields; but this exhausts the soil, and in older regions rotation of crops must be practiced, fertilizer must be used, and wheat raised less frequently. In small fields in the East broadcast sowing and

FIG. 7. Threshing on the prairies

the cradle are still in use, and on many small plots in parts of Italy and elsewhere in the older countries the soil is still prepared by hand and the grain is reaped with a sickle.

6. Flour milling. Wheat would form a nourishing food if threshed in the hand, winnowed with the breath, and ground without the aid of machinery. The wheat of modern times, however, forms food for nations through elaborate processes of manufacture by which a few skilled workers make it ready for the many. The story of the past shows a natural series of milling methods. In early days, as among rude peoples now, there was grinding by hand between two stones. The gristmill by the stream sends the grain between dressed millstones and "bolts" the flour, separating the middlings and the bran. The

[1] Writer in United States Census of 1880.

great merchant mill, however, crushes the wheat between rollers of iron or porcelain, and sends it in fact through several series of rollers, with an intermediate sifting or winnowing between successive grindings. Thus the best flour and the most of it is made by effectively separating the inner parts of the grain from the outer shell or covering. Hence we have the roller-process flour in brands which are known throughout the United States, and indeed in all lands where wheat serves as food. In the older process the "middlings" were discarded, but now are utilized to make the strongest and the distinctly high-grade or "patent" flour.

7. Milling centers. In the Old World the highest fame and skill in the making of flour have been reached by Hungarian millers at Budapest, the capital of Hungary. On the Danube, at the head of the great plains of Hungary, this was a natural place of manufacture for wheat products. There the roller process was used before 1840, and thence carried to other lands. The municipality whose greatest industry is milling has built on the Danube quay a great elevator in order to exhibit modern methods of handling grain.

In the first half of the last century Rochester, at the lower falls of the Genesee River, became the chief seat of the flouring industry of the United States. The wheat lands of western New York surrounded it, the falls furnished ample power for the mills, and the Erie Canal, on its opening in 1825, gave the needful carriage to the markets of the East.

Budapest sprang from a Roman camp, but the city of Minneapolis stands on ground where seventy years ago the Indian and the buffalo were roaming. In 1850 the few settlers sent their wheat three hundred miles to have it made into flour, but in that year a small feed and flour mill was built. In 1867 the town was made a city, and men who had grown rich in handling lumber were turning their attention to wheat. Both were products of the Northwest, and both found ready approach to Minneapolis. Arriving there, the power of the Mississippi River in the falls of St. Anthony was available for sawing and grinding. The first merchant flour mill was erected in 1854, and the bran was fed to the fishes of the river. After 1859 the industry grew, and the Hungarian processes were introduced in 1871. The Pillsbury mills are said to have enlarged their success by putting the keener American workmen to the use of the European methods. In 1901 the capacity of Minneapolis mills was 80,000 barrels per day, and one mill could turn out 15,000 barrels. Here again, on a larger scale than at Rochester, we have three essential things, — raw product, plentiful power, and ample transportation from the field or to the market by many lines of railway. Lumber and flour have created an industry in

FIG. 8. Flour mills of Minneapolis, St. Anthony Falls, and viaduct of the Great Northern Railway

barrels and boxes, and most naturally also the manufacture of breakfast cereals has grown to large proportions.

8. Domestic trade in wheat and flour. In 1886 Massachusetts raised 160,000 bushels of wheat. This would feed but a small part of her people at that time. In 1887 her product was 16,000 bushels, and for the past twenty years it has been too small to be reported. Much the same is true in all of New England. This means that nearly all the flour consumed by almost 6,000,000 people is brought from other states. New York in 1906 raised 9,350,000 bushels. As this state has about 9,000,000 people, and the annual consumption is five or six bushels to each person, it is apparent that most of the bread must come from the West. In early days each farmer tilled his field of wheat and had the grain made into flour at the grist-mill. Now if he raises wheat, he often sells it and buys patent flour from the great mills. Thus each part of the country is doing its special task, and exchanging its products with other sections. This is made possible through improved transportation, and thus has grown up a vast volume of internal commerce among states and cities.

We will take an example of such trade and follow the movements of a bushel of wheat, grown perhaps in the Red River valley. The farmer's wagon delivers it at the nearest railway

FIG. 9. A group of village elevators in the Northwest

station, where it is loaded on a car or is hoisted into a local grain elevator. It may pass by rail to Minneapolis, where it becomes flour. It may then be shipped by rail via Chicago to New York, Boston, or Baltimore, or it may pass by rail to Duluth, and by steamer to Buffalo, and thence by the Erie Canal or by rail to eastern centers. Again, the wheat may be shipped from the field in North Dakota, or Winnipeg to Duluth, and thence in a "whaleback" steamship to Buffalo and to local milling centers in the East, whence the flour is distributed to wholesale houses in the cities and to retailers everywhere. Great baking industries consume large consignments of flour, and the finished product, ready for the table, is in like manner carried by the railways and passed over to the wholesale and retail trade.

In the modern wheat industry the grain and flour are almost always handled by mechanical means. This is illustrated in the drill, the harvesting thresher, the railway, the steamship, and the grain elevator. The first grain elevator in the world was erected in Buffalo in 1843. Such buildings are conspicuous in all cities that handle grain. Endless chains hoist the grain from wagon, car, or steamship to any height, and drop it into bins according to kind or grade. Modern elevators are often built of steel and have facilities for drying or scouring the grain. Loading and reshipment are accomplished easily and speedily, and the transfer is made at small cost. A single "whaleback" may carry a quarter of a million bushels, and some elevators will store ten times that amount.

9. Financial importance of wheat. A product which is grown in large quantities and which hundreds of millions of people regard as a necessity has great and immediate value. This fact is important to the farmer who raises little but wheat, for with the season's returns he will pay off his mortgage, build a new home, or increase his equipment or his surplus. The local merchant in the nearest town instantly feels the result of a short crop or the advantage of a large yield or high price. The great city merchant in turn soon experiences the favorable or unfavorable effect on his business. Large shipments mean prosperity for railway and ship owners, for millers, employees, car and ship builders, coopers, lumbermen, rolling mills, machine shops and mines. There is money for the banks, and hence for enterprises of every character. The money required to "move" a single crop of wheat finds its way through many channels and affects an entire people. We thus can understand the interest with which crop reports are read in the growing season.

We understand also why wheat is in some respects the most important of commercial crops in the United States. The bulk of wheat is raised for immediate sale, in this respect differing from corn. Dondlinger [1] goes so far as to say that "directly or

---

[1] The Book of Wheat, p. 215.

indirectly it is the chief feature in our commercial relations." A strong foreign demand may increase the product and at the same time raise the price of bread at home.

Wheat, having long been a large export crop, tends to increase imports of gold, and thus to enlarge the holdings of the banks and promote home enterprises. Hence fluctuations in the size or price of a large export crop would have an important effect on home business.

It has been shown[1] that wheat is in a direct way more important in American commerce than corn or cotton, although these products may show larger total values. Corn is fed to stock and only a small part of it goes into export trade, as grain, though much of it finally does so as meat. Cotton exports are much heavier

[1] Andrews, Influence of Crops on Business in America, *Quar. Jour. Econ.*, Vol. XX, (1906), p. 323.

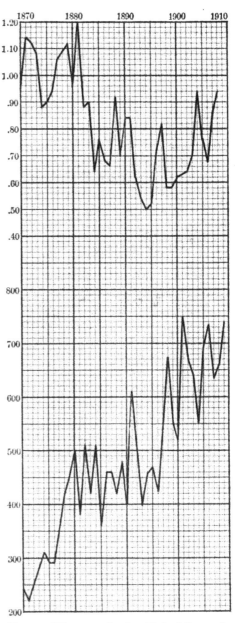

FIG. 10. Wheat production, United States, in millions of bushels, 1870–1909 (below); price per bushel, 1870–1908 (above)

than those of wheat, but cotton is not therefore the chief factor in our foreign trade balance. About two thirds of the crop is exported, with little variation, for the United States produces about three fourths of the world's cotton and is therefore little affected by the competition of other regions. The price is mainly controlled by the American supply. We grow, on the other hand, less than one fourth of the world's wheat crop. A large crop at home coinciding with a small crop abroad raises

Copyright, Kaufmann, Weimer and Fabry Company

FIG. 11. The "wheat pit," Chicago

prices, promotes export trade, and thus affects the international balance in an important way. A small crop at home with large crops abroad would, on the other hand, greatly reduce exports.

Cotton also is raised near the Atlantic seaboard, to which much of it goes on its way to market. Wheat is raised in the interior, and the railways have the advantage of long hauls. Hence Andrews concludes that the wheat crop is of "primary significance" to foreign trade, to the bank reserves of financial centers, and to the railway and shipping interests.

Large institutions like the Chicago Board of Trade furnish a center for the grain market of the country and serve a useful purpose, despite the speculation which goes with the volume of honest trade. The condition of wheat in any wheat land of the world is at once reflected on the local market. There is thus, in a sense, no local market for a grain of world-wide importance. The following paragraph gives a group of sample telegrams which might affect the price of wheat on the floor of a grain exchange.[1]

" The weather forecast promises rain in Kansas. The monsoon in India is overdue. Roads are bad in the Red River valley, preventing grain deliveries. The London *Times* has a cable that locusts have appeared in Argentina. A big mill in Minneapolis will shut down next week because the flour trade is dull. Navigation on the Danube will open unusually early. St. Louis has received fewer cars of wheat than on the corresponding day last year. Australasian grain, to arrive, is freely offered in Liverpool. There are rumors of strained relations between England and Russia in the Far East."

Thus wheat, by reason of easy shipment, swift communication, and the needs of many nations, becomes a major article of trade and is a commodity of the first order in the world of food and of commerce.

10. Wheat growing in Europe. The average wheat production of all Europe in the years 1901–1905 was 1,840,000,000 bushels. This is more than any other continent as yet produces, but when we remember that Europe contains about 400,-000,000 people we shall see why, in those years, an average of 250,000,000 bushels was brought in from other lands.

Some European countries raise little wheat and import much, while others raise a surplus. The reasons have to do with soil, climate, and with the density of population and the occupations of the people.

In the seven northwestern or Teutonic countries production is usually small and the amount brought in is large. This group

---

[1] Article in *Century Magazine*, 1903.

includes Great Britain, Germany, Belgium, Holland, Denmark, and the countries of Scandinavia. In the years 1901–1905 British farmers supplied but 21 per cent of the wheat needed in the United Kingdom. Fifteen years earlier they furnished 35 per cent. The reason for the change is that manufacturing developed and crops other than wheat were required by a growing population. At the same time new lands were opened in other parts of the world where wheat could be raised, and from which it could be transported more cheaply than it could be

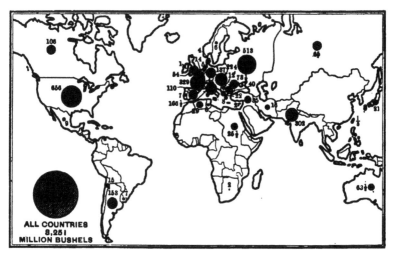

FIG. 12. Average annual production of wheat in millions of bushels, 1904–1908. New Zealand not shown, 7,000,000 bushels

grown at home. The chief British supplies were drawn from the United States, Russia, and from the British dependencies, Canada, India, and Australia.

Germany raises more of her breadstuffs, including much rye, but, like the United Kingdom, and for similar reasons, imports much more wheat than she did twenty years ago. Belgium in 1901–1905 produced but 23 per cent of her wheat, importing nearly 68,000,000 bushels. Norway raises but 10 per cent of her wheat. She is too far north and too rugged, and her small amount of arable land is required for other crops.

Southwestern Europe presents a different showing. Here are the Latin countries, France, Spain, Portugal, and Italy. Of these it may be said that they raise most of their wheat, but import a small percentage. The population is dense, but the climate is favorable, tillage is thorough, and labor is cheap. In the years 1901–1905 Latin Europe raised an average of 615,000,000 bushels, which was 91 per cent of the amount needed. All together these countries imported 60,000,000 bushels annually, or about what Minnesota or North Dakota might produce in one crop. France alone, of about twice the area of Colorado, raised 336,000,000 bushels, or 97 per cent of her needs. Italy produces 84 per cent of her consumption. She imports considerable amounts of hard wheat for the making of macaroni and semolina, and some of these is in turn exported.

Eastern Europe has a more scattered population and wider areas of suitable land, and hence produces a surplus. Among these countries, chiefly Slav, are Russia, Hungary, Roumania, Bulgaria, and Servia. The plains of the Danube in Hungary are prime ground for wheat, but most of the surplus goes to Austria and hence is retained within the empire. The great exporting country is Russia, which sent out in the years 1901–1905 an average of 141,000,000 bushels.

Russia is one of the great countries of the world's wheat industry. Wheat will grow almost anywhere south of the latitude of St. Petersburg, but it is raised chiefly in the "black soil" belt, which is in southern Russia, toward the Black Sea and along the lower Volga. To the north lies the industrial region about Moscow, whose winter climate is too severe for favorable growth. The dark soil is from a few inches to three or four feet deep, is, as its color would suggest, full of organic matter, and covers more than one fifth of European Russia, or 192,000,-000 acres. Nearly three fourths of the cereals of Russia are raised in this belt. The yield per acre is low, but the farmers have had little opportunity to learn modern ways, and transportation is poor. Thus Russia has not yet taken its full place in the competition of the world.

Some Russian wheat goes across central Europe to Baltic
ports. Siberian wheat largely seeks Archangel, and much goes
to the Moscow industrial region for domestic consumption. The

FIG. 13. Proportion of cultivated land under wheat in the several
governments of Russia in 1904

most important wheat lines, however, reach the Black Sea. The
run of most Russian wheats to the seaports is shorter than in
the United States. Odessa is one of the great ports, and Taganrog, on the Sea of Azov, is the largest port in the world for

the export of macaroni wheat. There is much inland transport by rivers and canals.

11. Wheat in other continents. This grain is raised in Siberia, in China, and particularly in India. Some provinces of India, although lying in a low latitude, have, by reason of their elevation, a climate resembling the summer in temperate latitudes. Wheat is raised on small farms and sometimes with the aid of irrigation. As in our own Northwest, the growth of wheat production has depended on means of carrying away the crop. These were provided by the Suez Canal and by the railroads which bring much of the grain to Bombay for export shipment. Great Britain alone in 1906 received 23,000,000 bushels of wheat from India. Much less would have gone to this market if ships had still to round the Cape of Good Hope.

In north Africa, Egypt and Algiers are important for wheat, and the latter, being a French colony, supplies to the governing land much wheat suited for macaroni and similar products.

If we follow the temperate zone in the southern hemisphere, we find three continental areas suited to wheat, — Argentina, South Africa, and Australia. Great Britain in 1906 received nearly 36,000,000 bushels from Argentina, almost as much as from the United States. Trade was so great, notwithstanding the long distance and the cost of transportation. If grain ships could be sure of return cargoes from Europe, freight rates would go down and Argentina would hamper the United States in the grain market of the Old World. A temperate climate and the wide plains between the Andes and the Atlantic favor grain production. Like Canada and Australia, Argentina is a country young in settlement and tillage, and is far from its limit of production.

In Australia, tillage is secondary to stock raising, rainfall is limited, and the yield per acre low. The quality of the wheat, however, is high, and in 1906 Great Britain received nearly 15,000,000 bushels of wheat from this source. South Australia and Victoria are the states in which it is raised.

12. Wheat exports of the United States. In 1850 the United States raised four and one-third bushels of wheat for each inhabitant. In 1900 the amount was 8.66 bushels per person. Allowing. 6 bushels each for home use, there was a large surplus for export. In 1906 our product was 735,000,000 bushels, and our population was probably about 88,000,000. Allowing 6 bushels each, we still have a large amount available for export, and during the last half of that year and the first half of 1907 there was sent out of the country 76,000,000 bushels of wheat and 15,000,000 barrels of flour.

As the country fills up, more land must be used for varied agriculture. There are also more persons to feed. Hence it has been urged that the United States may, at no distant time, raise no more wheat than will be needed at home, and that newer lands, like Canada, Argentina, and Australia, must supply the deficient countries. But the capacity of such a country as France, with intensive tillage, warrants the belief that the day of shortage is far in the future, and that exports will continue for many years.

13. General view of the world's wheat. We are now prepared to review in their order the stages through which this grain passes; namely, production, manufacture, transportation, and consumption.

14. Production. Wheat is a product of temperate latitudes, and this gives it a fairly definite geographic distribution, which man may slightly modify, as by irrigation and by breeding new varieties suited to shorter seasons or less rainfall. Plains are more suited to handling the crop than are hills and mountains, and have favored the invention of modern machinery, which in turn has made it more difficult to raise wheat with profit on lands with rugged surface or unsuitable climate.

15. Manufacture. Millstones have largely given way to roller processes and the small mill is replaced by the large. The industry is therefore concentrated in convenient places, such as Minneapolis, Budapest, or Bombay, and is conducted with special skill and on a large scale. At the same time the range of manufacture has widened from the making of various grades of flour, to include the alimentary pastes of Italy, France, and Switzerland, and the breakfast foods and biscuits of the United States.

16. Transportation. Movement of the product occurs between production and manufacture as well as between manufacture and consumption. Kansas wheat may be milled at Kansas City, St. Louis, or Baltimore; or it may cross the Atlantic to Liverpool, London, Rotterdam, or Hamburg, and be ground in foreign mills. It has recently been urged that we cannot afford to export whole wheat, because we lose labor, food for stock, fertilizer, and the special advantage of the American brand placed on flour made in this country. The transportation companies, however, favor the whole wheat by lower rates, because it is handled by elevator methods and thus more cheaply than flour.

The development of long-distance carriage deeply affects the areas of production, the price of the product, and the occupations of peoples. The railways of the Northwest and the steamships of the Great Lakes make possible the wheat harvests of that region. The profits give a wide range of desirable things to the grower, and the cheap transportation gives wheat bread to many among the poor of Europe to whom it was once an unavailable luxury. We have already seen how transportation made wheat a profitable crop in India. New facilities will increase its amount and importance in Russia, Canada, and Australia. The wheat lands of Argentina and Russia are nearer the sea than those of the United States, and this will favor them as competing grain producers. England is enabled to devote her energies to manufacture and commerce because transportation makes it possible to buy bread from remote fields. Thus specialization among nations is favored.

17. Consumption. Through this long series of processes the consumer is at last reached. He receives better bread, won at less cost to himself than in the former days when he raised his own wheat and prepared it for food by primitive methods. He is free to do other things, and the true aim of industry is reached, — the well-being of the individual. Meantime, like ends have been achieved by the many who have been instrumental in feeding him.

CHAPTER II

COTTON

18. The cotton plant. The usual height of the cotton plant is from three to five feet. It is perennial in regions which are free from frost, but in ordinary culture is treated as an annual, the

FIG. 14. Cotton bolls, unopened and opened

seed being planted each year and the crop gathered at the end of the season. The important part of the plant is the fruit or "boll," which follows the showy blossoms and is a pear-shaped seed vessel, which opens along several lines as it ripens, displaying a mass of fibers surrounding the seeds.

19. Varieties of cotton. There are many botanical species and these have developed varieties in great numbers, but for practical uses it is enough to note the chief cottons of commerce. These are the ordinary, or American upland cotton; the sea island, with longer, finer, and silkier fibers; the Egyptian; and the Peruvian, which differs in texture from other cottons and is often mixed with wool.

20. Climate and distribution. Cotton is regarded as a tropical growth, but the most important production is in subtropical regions intermediate between the more heated and the temperate zones. It requires at least six months without frost to mature its fruit, and this places a latitude limit, which, however, is very

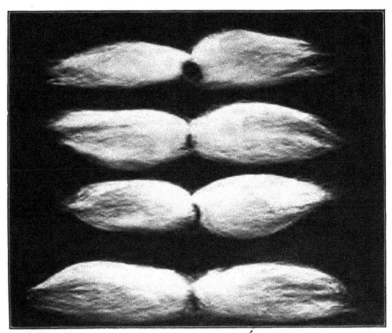

FIG. 15. Cotton fibers, attached to the seed

variable. Nearly all the cotton is raised between 40 degrees north latitude and 20 degrees south latitude. In the longitude of Europe it extends from the region about the Mediterranean to the Cape of Good Hope. In the Orient it reaches from Japan to northern Australia. In the western hemisphere we may place its limit in Virginia on the north and at Buenos Aires on the south. It requires a moderate amount of moisture during the months of germination and growth, and needs abundant sunshine during the stages of maturity and ripening.

21. Enemies of the cotton plant. This crop, like others, is subject to losses by unfavorable weather, as drought in the growing time, or rains in the maturing season ; but we here refer especially to pests, such as cutworms, which destroy the young shoots, and, in recent years, the boll weevil, which has been considerably destructive in parts of Texas, Louisiana, Arkansas, and Oklahoma. Here, as with many other crops, the United States Department of Agriculture has investigated methods of selection and of culture which aid in checking the evil, and thus this

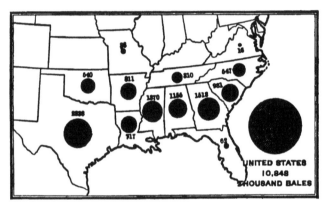

FIG. 16. Cotton, — average annual production in thousands of bales, 1899–1908

branch of the national service becomes an important factor in the geography of commercial products.

22. Cotton production in the United States. Cotton was first grown here in Delaware and in Virginia. Only Virginia and the Carolinas raised cotton to any extent in the eighteenth century. In the latter half of that period interest grew because cotton machinery was invented in England, and production reached southward and westward. In 1850 the American cotton center was near Birmingham, Alabama. It has moved steadily westward since that time, and in 1899 Texas raised one fourth of the national crop. For one year (1906) more cotton was raised west than east of the Mississippi River. This was especially due to increase in Texas and Oklahoma.

Cotton is grown northward to a line reaching from Old Point Comfort, in Virginia, to Cairo, at the mouth of the Ohio River. Early settlers north of the Ohio River raised some cotton for home use, but ceased when railways made exchange possible, affording an illustration of the tendency to specialize in agriculture as a region matures. Since 1850 the South has had more land in corn than in cotton, but the ratio changed as railways began to bring corn from the prairies. The end of the last century brought in a diversified agriculture, and corn has now greatly increased in acreage, though cotton remains the great money crop.

The cotton crop of 1904 was one of the largest to the present date, consisting of nearly 7,000,000,000 pounds. The value of the crop in 1907 was $700,000,000 and the acreage was equivalent to nearly 49,000 square miles, — an area of actual cotton fields five sixths as large as all England. For the years 1899–1907 the relative rank of the great cotton states was as follows : Texas was first every year; Georgia and Mississippi were rivals for the second place ; Alabama came next, — all of these being Gulf States. South Carolina surpassed Louisiana, and the remaining Gulf state, Florida, stood eleventh. Virginia, one of the original cotton states, had dropped to thirteenth, and Oklahoma ranged from sixth to tenth in position.

FIG. 17.. Percentage of United States cotton grown in each state, 1909

23. Cotton in other countries. Cotton is believed to have been raised in India longer than in any other land. Records tell of its use several centuries before the Christian era, — two thousand years before it became an article of trade in Europe. Herodotus refers to it as "tree wool," while in modern times Columbus found the plant in the West Indies, and Cortes discovered it in Mexico. Manchester cotton buyers are said to have been in the Levant about 1640, and manufactured cotton was sent from Bombay to Great Britain in 1666. India is still one of the important exporting countries, but considering the great population and the demand on the land for other crops, it is thought that her export of cotton may not be greatly increased.

China raises much cotton, but it is largely used in the homes of the people, and the residue which enters commercial channels is not an important part of the world's supply. The Russian possessions of Turkestan and Transcaucasia rank next among Asiatic countries. Here production has grown through the extension of railways, and has been encouraged by the Russian

FIG. 18. Cotton, — average production in thousands of bales, 1903–1907

government by means of duties on imported cotton and in other ways. Small amounts are raised in Turkey, Persia, Korea, and Japan. In Africa, Egypt is the greatest cotton producer, and the country ranks third as regards the world's factory supply. There is no danger from frost, the growing period is long, and the country is free from disastrous storms. There is abundant water when the plant needs it, and labor is cheap. Here, as in the Southern states of America, it is the money crop, and has been called the "backbone" of Egyptian agriculture. The fiber grown along the Nile is second only to the sea-island product in fineness, silkiness, and length, and these qualities create a strong foreign demand.

Of South American countries Brazil and Peru require notice. For more than a century Brazil has exported cotton, especially to Great Britain, but the amount has decreased in recent years and the production is more important for home supply than for export. The culture is of long standing in Peru, where cotton fabrics have been found in ancient tombs. It is grown by aid of irrigation in the coast valleys, and is known as " rough Peruvian," by reason of a woolly character which fits it for mixing with wool, a fact which creates a steady demand for it in the United States. In a small way cotton production reaches over into Argentina.

In 1786 the West Indies produced 70 per cent of the British supply of cotton while the United States and India combined furnished less than 1 per cent. Cotton is not now an important crop in the West Indies. The United States, India, and Egypt are the three great commercial countries in this product, the first named having, in 1907, grown 66 per cent of the commercial cotton of all lands. Wheat was an important international commodity many centuries before this became true of cotton. In both cases the centers of production have shifted rapidly as new lands have opened or old lands have come under enlightened administration. Thus we mark a progressive adjustment to environment, while the supply keeps pace with growing population and with the more elaborate requirements of recent times.

FIG. 19. Percentage of world's mill supply of cotton contributed by each country, 1909

24. Distribution influenced by special conditions. In 1869, encouraged by the high prices then paid, farmers in southern Illinois raised 200,000 pounds of cotton, and smaller amounts were grown in other states north of the cotton belt. On the decline in price these farmers went back to other crops. The American Civil War (1861–1865) nearly cut off the supply from the southern states and promoted cotton culture in other countries. Egypt and India were among the regions which profited,

but Egypt is the only land which did not fall back to its old status when American cotton resumed its place in the world's market. Russia has promoted cotton culture in her Asiatic provinces, and the republic of Colombia is paying a bounty on native cotton, hoping to establish the industry in suitable territory as large as the state of New York. In like manner the British Cotton-Growing Association is promoting culture in Africa, Sind, and the West Indies.

In 1900 our direct imports of cotton from Egypt were valued at $16,000,000. The sea-island product does not give us enough of the fine, long-staple cotton for our home use. The Bureau of Plant Industry of the Department of Agriculture is growing Egyptian cotton in the region of the lower Colorado River in Arizona and California. The summers are dry and long, the soil is alluvial and deep, and irrigation is practiced. These are conditions close to

FIG. 20. Areas reporting sea-island cotton, 1909

those of Egypt. About 600,000 acres will soon be "under the ditch," and it is believed that one fifth of this area would raise as much of this long-staple fiber as we must now get from Egypt. Thus the interest of nations or of particular groups of men aids in extending or changing the distribution of a product of the soil. The conditions of transportation also have their bearing. For example, along the lower Colorado the available market and the cost of carriage must be factors in the problem, as well as the scarcity and high price of labor.

25. Cotton from the seed to the bale. In the cotton states the seed is usually planted in April. This was done by hand until planting machines were invented. Much attention is paid by progressive growers to the selection of the seed, so that the greatest yield, or highest quality of "lint," may be produced. After a period of careful cultivation the picking begins, usually in August, and lasts through the autumn. This is the one remaining process which is still accomplished by hand, and it is

FIG. 21. Cotton picking

therefore costly. Pickers go over the field at intervals, gathering the bolls which are ripe.

The lint is separated from the seeds by the process of ginning, and the great development of cotton in the United States dates from 1793, when Eli Whitney, an American, invented the saw gin. It has been said by a famous English writer that Whitney thus advanced the United States in as high degree as did Peter the Great the empire of Russia. The gin made it possible for more people to use cotton, and profitable for more farmers to grow it. Hence we have here a labor-saving machine which has affected the industry and the customs of wide areas of the globe, and which illustrates an important aspect of geography. There are about 30,000 ginneries in the United States, and the tendency is to abandon the smaller private plants and concentrate in large public ginneries. After ginning, the cotton is pressed into bales of about 500 pounds each, and is then ready for transportation and manufacture.

26. Cotton manufacture in the United States. The cotton of the South goes to local mills, to those of the North and West, or

FIG. 22. Fulton bag and cotton mills, Atlanta

is sent abroad. New England has been the historic center of manufacture in the United States, partly by reason of her ample water power, and partly because of the enterprise and skill of her people. The cotton states were in early times almost wholly agricultural. During the past thirty years, however, there has been an industrial awakening in the South, and much of the cotton is now manufactured within short distances of the fields in which it has grown. This is favored by the abundance of water power and coal found in the region of the southern Appalachians. New England, on the other hand, has gone beyond the limits of her water power at some of her mill centers, such as Fall River and Lowell, and must haul coal. These geographic conditions tend to restrict cotton manufacture in the one region and favor it in the other.

In 1907 North Carolina led all other states in the number of cotton mills, having 276, while Massachusetts had 204. Massachusetts, however, had mills of larger capacity and consumed nearly twice as much cotton as her Southern rival.

FIG. 23. Cotton levee, New Orleans

Georgia followed, with 149 mills, and then came South Carolina, Pennsylvania, and New York. Comparing the cotton states and all New England, in 1880 the former consumed but one sixth as much as the latter. After thirty years, in 1910, the cotton-growing states used 2,292,000 bales, and New England 2,016,000 bales. New England, however, continues to operate many more spindles than the cotton states, and has more capital invested in cotton manufacture. The finer grades of cotton are still in large part made in the Northern mills.

Brooks and Doxey, Manchester

FIG. 24. Spinning room of Lancashire cotton mill

32

It seems clear that the United States, with its proven industrial capacity, should not be content to export so large a share of its raw cotton. This ought rather to be manufactured in American mills and the finished goods sold directly in foreign markets, giving full advantage to American labor and capital.

27. Export and import trade. The first export of cotton from this country was by Virginia, from which eight bags were sent in 1784. In 1790 England received less than one sixth of one per cent of her cotton from the United States. In recent years the United States has supplied about two thirds of the cotton imported by all nations. Considering the disposition of a single crop, — that of 1907, — we find that 33 per cent was consumed at home, 57 per cent was exported, and 10 per cent was left over at the end of the year.

In a single recent year only 14 per cent of our export cotton left the country by ports outside the cotton states. The three chief ports, in the order given, were Galveston, New Orleans, and Savannah. The first exported more than the two others combined. The destination of our export cotton shows interesting changes. In 1880 Great Britain took two thirds of it, France one tenth, and Germany one twelfth. In 1906 Great Britain received four ninths, Germany one fourth, and France one ninth. Liverpool and Bremen are the largest ports of entry for our cotton, while considerable goes to Genoa and other Mediterranean ports. The skill of American manufacture has also made itself felt in foreign trade, and in China, for example, American shirtings are said to have displaced in large degree the native fabrics. The United States imports raw cotton chiefly from Egypt and Peru, to secure fibers of special kinds not sufficiently produced at home.

28. Foreign manufacture of cotton. In the latter half of the eighteenth century a series of important inventions gave a strong impulse to the spinning and weaving of cotton in Great Britain. Inventive genius and capacity for business combined to give primacy in this industry to the British people. The central position and good harbors of the kingdom favored the expansion of trade, and large supplies of coal and iron were at hand. These gave power for mills, locomotives, and steamships, and the material for mill machinery, railways, and modern ocean craft. Growing population made labor plentiful and cheap, and the raw material came readily to the doors from expanding American plantations. Ample markets were found near at hand on the continent, and among almost all nations across the seas.

In 1909 there were in the world 133,000,000 cotton spindles. Of these the United Kingdom had 53,000,000, the United States 28,000,000, Germany 10,000,000, Russia 8,000,000, and France 7,000,000. The number of spindles, however, does not show the relative amounts of raw cotton consumed, for in 1898 the United States passed the United Kingdom in this regard, and in the year 1906–1907 used over 5,000,000 bales, as compared with a little over 3,500,000 consumed by the rival nation. Cotton spinning has considerable importance in India, thus limiting the British market somewhat in the

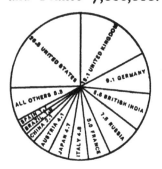

FIG. 25. Relative consumption of the world's cotton by manufacturing countries for year ending August 31, 1909

East. The advance of cotton milling on the continent curtails the market there, and some of the more modern machinery permits certain grades of work to be done by the less skilled and cheaper labor of Asia and of the southern states. We are thus able to see that the growth and stability of an industry depend on many factors, some of which are geographic. Among the conditions are business sense ; inventive skill ; availability of raw material, either in the field of production or by favorable transportation ; the supply of labor, of motive power, and structural material ; and a market which cannot be seized by those having greater advantages. To this last might be added government regulation of trade and general commercial and financial conditions, such as the supply of capital and the state of trade at home and abroad.

29. Cotton-seed products. The cotton plant is primarily grown for its fiber. The seed has been adequately used only during the past generation. Before that time it was sometimes employed as a fertilizer, but often went to waste and in a manner that menaced health. The seed now forms the basis of a vast business, and may be taken as a good type of the by-product in modern industry. The making of oil from the seeds began in England and became commercially important in the United States following the year 1870. In 1890 there were in

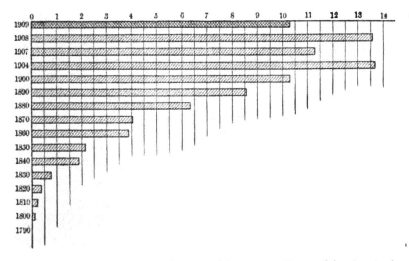

FIG. 26. Cotton production of the United States in millions of five-hundred-pound bales, 1790–1909

this country 119 mills, and in 1907 the number had grown to 786. In 1899 the total value of the cotton seed was almost $50,000,000, about one eighth of the value of the entire cotton crop; and in 1905 the value of all cotton-seed products was $96,000,000.

One ton of seed produces 40 gallons of oil and 700 pounds of oil cake. The oil has largely displaced olive oil in southern Europe, being used both under its own name and as an adulterant. It is used in making substitutes for lard and butter, and in the manufacture of soap, and is to a small extent employed as a lubricant. The oil cake serves for fertilizing, and is said to be as useful for this purpose as if the oil had not been removed. Its return to the soil, either directly or through serving as food for stock, is important, because one crop of seed exhausts the soil as much as ten crops of fiber. The oil cake and meal are bought for Danish and German dairies. When one buys the famous Danish butter in an English city, he is still, therefore, depending in a measure on the agricultural resources of the United States, tracing the butter to the same source from which the loaf of bread may also have come.

The leading foreign customers for the oil in recent years have been France, Germany, the Netherlands, the United Kingdom, and Mexico. Hull, on the eastern coast of England, has nearly twenty mills, and is still the largest crushing center, bringing most of the seed from Egypt and a small amount from the United States. Memphis is the rival of Hull in this industry.

The hulls of the seeds are also valuable, both as food for stock and for the making of paper, — a phase of the paper industry for which several mills have been established. In like manner the stalks may be cut and used as a fertilizer, they may be fed to stock, or the fiber may be used for bagging. We see strongly emphasized here the tendency, everywhere present in modern industry, to let nothing go to waste. This end is promoted primarily by the desire for increased profits. It is made possible through greater technical skill, especially in chemistry and in mechanical invention, and it results in a more complete use of the resources of the earth and in improved conditions of living.

30. Uses and relations of the cotton plant. Some plants, such as the coconut palm, or an animal like the reindeer, may furnish almost the entire round of necessities to the people among whom they are found. This could not be true of any product among advanced nations having varied wants, but the cotton plant nevertheless offers·an impressive variety of uses. These uses are exhibited diagrammatically in *Bulletin 95* of the Twelfth Census, and the census table is reproduced on page 37.

It should be explained that "linters" are short fibers remaining after ginning, and recovered from the seed before the oil is pressed out. A study of the table shows the following uses of cotton: clothing, various fabrics, surgery, medicine, explosives, paper, food for man, food for stock, lubricating oils, fertilizers, soaps, fuel; and we may add the cylinders of phonographs, made from the residue in the refining of oil.

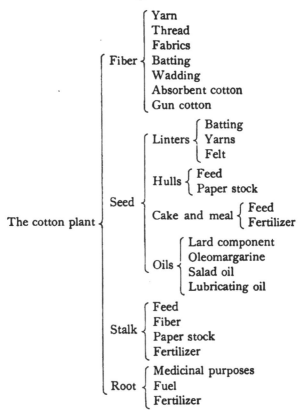

We may now observe that all the processes required by cotton planting, cultivating, ginning, transportation, milling, and marketing make demands on the forest and the mine, and establish great communities. These communities in turn make corresponding demands upon the fields near and far, upon the mine, quarry, and forest, and thus actions and reactions are set up throughout the commercial world. The plantation, the mill, the wholesale and retail store, and the consumer are linked together by lines of communication which reach all nations that have risen above the primitive stage.

31. Use of cotton in comparison with wool and flax. The statements in this and the following section are taken in substance from a paper on the future demand for American cotton, in the " Yearbook of the Department of Agriculture " for 1901. In 1800 wool and flax were in nearly equal use in Great Britain, flax being a little in the lead. The amount of cotton consumed was less than half the amount of either. Sixty years later, at the time of the American Civil War, wool was somewhat in excess of flax, and cotton was more used than wool and flax combined. In 1900 wool was two and a half times as important as flax, and cotton exceeded both the others twofold. In a century flax doubled, wool was multiplied by five, and cotton increased thirty-nine fold. An Irish journal a few years ago discussed the ascendency of cotton in an article on the " Decline and Fall of the Linen Shirt." Cotton also has become much used as a substitute for woolen, and in cotton and wool mixtures. Even silk has met competition, for many purposes, in mercerized cottons. Some silks of foreign manufacture are said to be heavily weighted and so mixed with mercerized cotton that it is sometimes difficult to find the silk. It is possible, in a great variety of so-called silk fabrics, to replace the silk in part with cotton, as in draperies, hangings, linings, and even in neckwear and hosiery. The two factors which have cheapened clothing for the world are the invention of cotton machinery and the opening of the cotton fields of the United States.

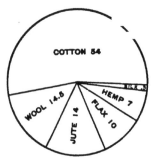

FIG. 27. Relative importance in quantity of the world's six leading textile fibers

32. The future of cotton. The population of the world is estimated at 1,500,000,000. Of these about one third are fully clothed, one half rank as partly clothed, and the remainder are nearly without clothing. With the expansion of civilization it may be expected that all human beings will ultimately require clothing, and this applies not only to the existing population but to such increase as the coming generations promise. As cotton is the cheapest of the great textiles and the most suited to the vast populations of warm climates, it will be in growing demand. There is little prospect of effective or large competition with the United States in growing cotton. It is believed that the Southern and Southwestern States can largely meet future requirements ; that Texas alone could raise 10,000,000 bales, an amount nearly equal to our entire present crop ; and that our total capacity may not fall below 25,000,000 bales. Whether or not cotton be "king," its commercial importance seems assured to coming generations.

CHAPTER III

CATTLE INDUSTRIES

33. General growth in the United States. If we go back two or three generations, we find each farmer keeping a few cattle, milking his cows, and fattening an occasional steer for home use. If near a town, the farmer might vend the milk from door to door, and the local butcher would purchase his surplus animals, which not uncommonly were cows no longer profitable for milk. Butter and cheese were made in primitive ways for domestic use, and the products of cattle raising could hardly be called commercial.

The cattle of those days were descended from ancestors that had been brought by the first settlers from England, Holland, Denmark, or other countries of northwestern Europe. At the same time the progenitors of Texas cattle had come over from Mexico, whither the breed had been carried by the Spaniard. These so-called native cattle began nearly one hundred years ago to be improved by crossing with pure breeds imported from the best modern herds of Europe. This advance has been marked since the middle of the last century, and at the same time beef, milk, butter, and cheese have taken their places among the articles of home and foreign commerce. The ends gained by higher breeding have been prolonged flow of milk, more of it, and higher quality. One cow often produces more than three or four animals of sixty years ago, with great saving in food, housing, and labor.

34. Distribution of cattle in the United States. The general farmer in the East still keeps one or more cows for home supply, and may slaughter a fattened animal for home packing and for fresh beef in the cold months. But in some sections, as southern and central New York, dairying is the chief industry of the farm, and herds of improved breeds are kept with a precise care and skill unknown a half century ago. Fresh pasturage, the silo, and corn from the Central States form the food, and the herds are kept for milk and its products, while fattening for beef is a minor element.

On the prairies and throughout the corn belt of the Mississippi Valley both the dairy and the beef trade have grown to large proportions and respond to Eastern markets and the foreign

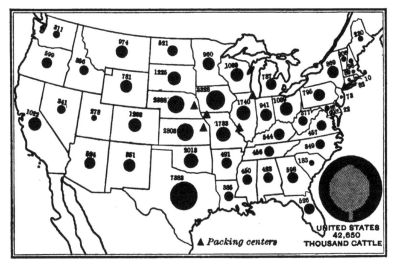

FIG. 28. Cattle other than milch cows; average annual number in thousands, 1899–1908

demand. Beyond the corn belt are the Great Plains and Cordilleran plateaus, whose cattle industry will be described in the next section.

Cattle raising is extending into the Gulf States and is locating in the neighborhood of the cotton-seed industry. This meets the double need of food for the cattle and fertilizer for the land, and is a good example of the causal relations which industries sustain toward each other. Dairying is extending on the Pacific coast, to supply the increasing population, and in the expectation of shipping products to the Orient, in which beginnings have been made. The Canadian government has promoted the dairy interests of the Dominion by a guaranty, to Pacific steamship lines, of payment for the use of ship refrigerators to full capacity, even though present exports fail to fill them. Apart from the milk supply for the cities of the East, the cattle region of the United States is the plain between the Appalachians and the Rocky Mountains, belonging chiefly to the basin of the Mississippi River. As we should expect, this industry is more independent of latitude lines than is wheat or cotton.

35. Ranges and ranches of the Great Plains and Cordilleras. Cattle raising is the foremost industry of the Southwest, a region including much of Colorado, over half of Texas, and nearly all

FIG. 29. Roping calves for branding on the western plains

of New Mexico and Arizona. The cattle region extends northward through Kansas and Nebraska, into the Dakotas, Wyoming, and Montana. These states of the Northwest are as good for grass as the more southern district, but the severe winters offer some hindrance to the cattle industry, particularly in raising the calves. In some regions, as in New Mexico and Arizona, there is much "free range," or open government land, occupied at will by herd owners, with their cowboys and horses, and the occasional roundups for branding the calves and securing the steers for market.

In Texas much of the land is leased from the state and fenced. In one case a single pasture fence runs 200 miles in one direction. On government ranges fencing is illegal, though the law is often violated. From 15 to 25 acres of the thin pasturage of the plains is required for each head of cattle. Overgrazing leads to the destruction of the grasses and the washing of the soil; hence leaseholders and owners of definite tracts are likely, for self-protection, to restrict the size of their herds. Free ranging is also narrowed by the progress of settlement. The best lands are taken up for tillage or as private ranches for the raising of cattle.

Thus cattle raising on the plains is becoming a settled business, conducted upon newer and more conservative methods. As an example of the scale on which the industry is carried on, a single ranch in the Southwest covers 5000 square miles and has 125,000 cattle, which are attended by 125 men and 1600 horses. Ten trains may be loaded for market from a single ranch. Improvement in breeds is now sought on the ranch as well as in the select herds in the Central and Eastern States. Irrigation is practiced, and dry-land crops, such as alfalfa, Kafir corn, and millet, are raised for the stock. It is now customary to "finish off" the ranch steers with a few months of corn feeding. Thus only can beef be produced which meets the demand of the best markets. About 75 bushels of corn must be fed to a single animal to bring him to prime condition. Ranch or range cattle are often sold to stockmen in the corn belt for this final stage of their preparation for market.

The grazing of cattle is permitted in all the government forest reserves without charge, but only by a permit which limits the number of animals and also the time. The herdsmen are required to be careful of fires, not to bed beyond a certain time in one place, nor within 500 yards of a stream or spring. The herds return to the prairie ranges in the winter, somewhat as Swiss or Norwegian cattle come down to the valleys at the end of each summer.

The cattle industry thrives for different reasons in various sections. In the East the chief reason is the presence of great cities, which need dairy supplies. In the Central States corn is the leading factor. On the plains, grasslands of great extent,

FIG. 30. The stockyards, Chicago

secured free or at small cost, offer the principal encouragement. In the East again the dairy is the most important ; in the West beef is the object ; and in the intermediate region both beef and dairy interests are large.

36. Centers of the packing industry. The chief western markets for live cattle are Kansas City, St. Joseph, Omaha, Sioux City, St. Paul, East St. Louis, and Chicago. The majority of the great packing establishments are in these cities. The work would naturally be carried on in cities of some size, to secure ample transportation for gathering the animals and distributing the product. These particular cities are suitable because they are in the corn belt, where stock interests are large, and because they are on the way from the ranges of the West to the markets of the East and of Europe. Minneapolis shows the same conditions in relation to wheat and lumber.

The Chicago Union Stockyards were organized in 1865, and the packing business grew strong because the refrigerator car was soon invented and began to move on the railroads. Packing could not be profitable if carried on only during four cold months of our Northern winter. Processes of refrigeration and other means of preservation have made the packer independent of changes of season or of latitude, and have given fresh meats to people who are far from the sources of supply.

The following facts about packing apply to other animals than cattle, chiefly sheep and hogs. The growth of the industry may be seen by comparing the whole value of the products in 1850 ($12,000,000) with the total in 1905 ($913,000,000). The business increased threefold in the twenty-five years following 1880. Of the total for 1905, more than half, or $505,000,000, belongs to the Central States, and chiefly to the cities named above. Packing is, however, growing in the southern states, where it increased 118 per cent from 1900 to 1905. This is a case of increasing diversity of agriculture in the older parts of the country as the stock areas of the newer West are narrowed.

In the year 1905 the state of Illinois made meat products to the value of over $300,000,000, or one third of the total for the entire country. Kansas stood second among the states and New York third, followed by Nebraska and Missouri. The single city of Chicago turned out products to the value of $269,000,000. In order of cities, Kansas City (Kansas) was second, Omaha third, the Manhattan and Bronx boroughs of New York fourth, and St. Joseph fifth.

The products from hogs somewhat exceed those from cattle, but fresh beef showed the largest percentage of increase from 1900 to 1905 of any of the chief meat products. The number of cattle slaughtered in the United States in the year 1905 was over 7,000,000.

37. Transportation ; methods of preservation. The refrigerator car was invented in 1868, and is now a familiar object on all leading lines of railway. The car, combined with depots for cold storage in the chief towns, affords a complete system for transferring meats from the place of slaughter to the market and the consumer. Thus Eastern cities need no longer ship beef " on the hoof," which often meant deterioration in quality. The transfer of live animals, however, is now carried on successfully even from the interior to the ports of Europe, the animals being well fed, regularly watered, and their well-being safeguarded by the law.

Refrigeration is effective through the country by reason of the new industry of manufacturing ice. This has grown greatly since 1870, and in 1905 there were more than 1300 plants, not including those in breweries and cold-storage warehouses. In the thirty-five years of the period named, ice declined from $20 or $30 per ton to a little more than $3 for the same amount. Ice making is not confined to warm climates, but is carried on in many of our Northern states as well, — Pennsylvania, New York, and Ohio being among the first. The ice has also been essential to the shipping of Southern fruits, and ice-making plants are often installed on ocean steamships. Here is a factor, absolutely essential to the meat industry, which has no place in the marketing of the commodities already described, wheat and cotton.

Meats are often fitted for keeping and carriage by being cooked and hermetically sealed. This method was first practiced in 1879, and canned meats figure largely in general trade, in army supplies, and in equipping expeditions of various kinds. Salting, smoking, and drying are other and familiar means of preserving flesh.

38. By-products of packing. In earlier times cattle were slaughtered almost wholly for the meat, the hides, and the tallow. Now, however, every part of the animal is used, the refuse being turned into fertilizers. The following objects have been named as ·a small part of the by-products of a Chicago packing house [1]: glycerin, neat's-foot oil, buttons, felt, bristles, soap, glue, pipestems, chessmen, knife handles, fertilizer, meat meal for poultry, brewer's isinglass, curled hair, gelatin, glycerol rennet. The horns serve not only for buttons, but for brush backs, combs, hairpins, and druggists' scoops. The shin bone, which is hard and strong, serves for the handles of toothbrushes, knives, and razors. Several obscure glands of cattle and other animals are made into medical preparations, but even with all these the list is far from complete. Meat packing is an example of industries in which modern invention has evolved a large number of by-products of great total value, furnishing often the major element in the profits. In lesser degree this is true of cotton, and another conspicuous example is petroleum.

39. Beef exports of the United States. Most of the live cattle exported from the United States are taken by Great Britain, which also is the best market for dressed beef, and buys large amounts of the canned meats. To avoid the propagation of disease, Great Britain requires all live cattle from the United States to be slaughtered upon their arrival. France and Germany are the next best customers for American meats. Japan is also a large importer of our canned beef. Most beef for foreign shipment goes by the port of New York.

Soon after 1880 there was municipal inspection of animals and meat at Chicago and Cincinnati, and in 1890 federal

[1] See *Century Magazine*, Vol. LXV (1902), pp. 148–158.

inspection was inaugurated by law. This law was enacted primarily because European governments were placing restrictions upon our exported meats. Legislation has since been extended, and the law of 1906 has provided, from that date, for most stringent inspection for the protection of both domestic and foreign consumers. This, however, applies to the large packing-houses and to interstate commerce. There is still need of municipal inspection, and there is danger from diseased animals and unsanitary slaughtering where " home beef " is sold.

40. Foreign producers of beef. Canada sends many live cattle to Great Britain, but has no considerable trade in dressed beef

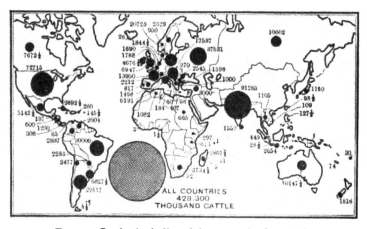

FIG. 31. Cattle, including dairy cows, in thousands

and canned products. Cattle are raised mainly in the eastern provinces, but the industry may extend, as in the United States, to the great central and western plains. Argentina is the other large producer of beef in the western hemisphere. In much of that country the grass grows throughout the year and stock never needs shelter. Steers sometimes have dry feed, or hay in dry seasons, and the need of final fattening with corn is now recognized. On the whole, production is cheaper than in the United States, and the country is therefore a strong competitor in the markets of Europe. There were about 28,000,000 cattle in Argentina in 1904, and the raising of blooded stock is now common.

Both frozen and chilled meats are exported, chiefly to England. Several freezing-plants with large capital have been established near Buenos Aires. There is effort at the present time to send out more chilled beef, since the frozen beef reaches the cheaper trade.

Australia and New Zealand are large producers of beef. In both countries the frozen-beef trade is considerable, but is surpassed by dairy products on the one hand and by the trade in mutton on the other. The state of Queensland is the Australian center of the cattle industry.

As in the case of wheat, European countries raise many cattle and supply some of their own beef. They use oxen for draft animals to a degree unknown in America. Russia, Denmark, and some other countries of Europe export meat, but the United States, Canada, Argentina, Australia, and New Zealand are the countries whose surplus supplies the deficiencies of other lands.

In the United States there is one milch cow to every four persons, and the yearly value of the products of dairying is about $500,000,000. The valuation has been increased more than fivefold within twenty-five years. At the same time the number of dairy plants and workers has about doubled, showing that the establishments are larger and that labor-saving machinery has aided in concentrating and handling the work.

41. Milk supply of cities. Down to 1850 the railways were not used to carry milk. All cities and towns were supplied by suburban dairymen, who owned and milked the cows and brought the milk in carts to the doors of their patrons. This custom still prevails in the towns and smaller cities, but in cities of the first order a great and complicated industry has grown up. Under the old plan the agriculture of the region adjoining a city was affected only in a local way. Now the ground required for dairy herds may reach out hundreds of miles, making other agricultural pursuits secondary and affecting transportation in large measure. As many large cities are in the East, milk is the chief product of the dairy farms, and the butter and cheese industries have been, to a considerable degree, driven westward to the Central States.

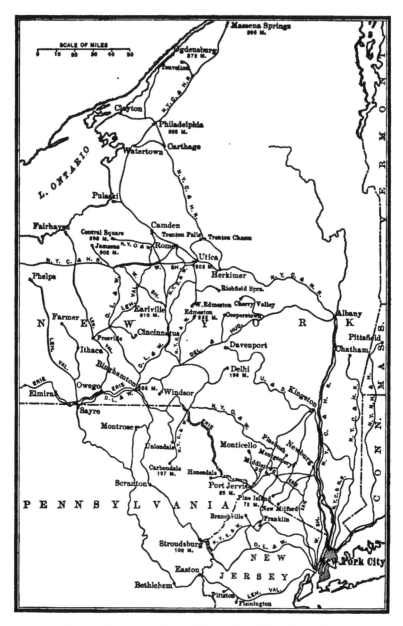

FIG. 32. Sources of the milk supply of New York City

New York illustrates the milk industry that is tributary to any great city. More than 1,000,000 quarts are consumed each day. If the cream used is included and added on the basis of the milk required to furnish it, the consumption rises to 1,500,000 quarts daily. Eighty-seven per cent of this supply comes from the state of New York, and the remainder from the nearer parts of New Jersey, Pennsylvania, Connecticut, and Massachusetts. The map of the milk routes supplying the metropolis shows a network of railways reaching out chiefly to the northwest as far as Lake Ontario and the St. Lawrence River. Thus the more distant sources of supply are three hundred to four hundred miles away. The New York Central, the Delaware, Lackawanna and Western, the Ontario and Western, the Erie and Lehigh Valley railways offer the chief means of transportation. Milk is received from creameries at many stations not used for passengers or general freight. The stations receiving milk are more than 500 in number.

The milk is brought to the creamery by the farmer or by a neighborhood collector, and here it is cooled and made ready for shipment. The cars, cooled with ice, are attached to local passenger trains, or in many cases the milk train carries a single passenger coach. At suitable points the milk cars are made up into exclusive milk trains, which are said to make the fastest time on the various railways. The great cities of modern times could not exist without transportation which is regular and swift. The vast population of London or New York could not live upon the products of their own limited neighborhoods. Milk, fish, fruits, and vegetables are all perishable, and must be used soon after their production.

It should also be noted that sanitation is more and more made a matter of public and private requirement. This applies to the health of the herd, the condition of stables, the character of the food, and the cleanliness of all vessels used in transportation. The state agricultural departments and the city boards of health have jurisdiction, and many of the milk-handling companies enforce rules which tend to secure milk and cream in perfect condition.

42. Butter. In the first half of the last century butter was made on the farm, in simple ways, by the members of the farmer's family. There were well-known producing regions,

FIG. 33. Sources of the milk supply in Boston

and "Franklin County," "Orange County," or "Goshen" butter was recognized as the best, in New York and Boston. In 1861 a creamery was established in Orange County, New York, and the factory plan, growing rapidly, became known in other countries as the "American system of associated diarying."

There was great progress in inventive appliances during the century. For seventy years the United States Patent Office issued a patent for a new churn on the average every ten or twelve days. A most important invention is the mechanical separator, by which the cream is removed from the rest of the milk soon after the milk is drawn from the cow. It is thus more sanitary than if left to " rise " in the primitive way, and saves the large expense

FIG. 34. City inspection of milk

of hauling the milk. The residue is left at home, and is serviceable as food for swine.

Another invention of the first order is an appliance for testing the amount of butter fat contained in milk, which thus enables the farmer to sell his product at its real value for butter making, and to realize any advantage which he may deserve from the high quality of his herd. In selling cream to the factory the farmer supplies the raw material of manufacture as truly as if he were selling wheat to the miller or bales of cotton to the spinner and weaver.

As in cotton production the picking is the only process in which little advance has as yet seemed possible, so in the dairy the milking is still left to hand work. In all other respects there is mechanical aid and the business is highly organized. In most states there are dairy associations, and the experiment stations and the national Department of Agriculture are ceaselessly active. There is thus much incentive to produce the best herds and to use the most advanced methods in all processes. Many journals give

FIG. 35. Bacteriological laboratory, State Dairy School, Ames, Iowa

useful information of appliances and markets, and dairy schools give suitable training to farmers and factory superintendents.

Among the states Iowa is the greatest producer of butter, having about 800 creameries and turning out one tenth of the national output. New York is next in rank, notwithstanding the inroads of the milk trade upon her dairy capacity. That capacity has been enlarged by surrendering other crops, as, for example, by substituting herds for hops in the counties of central New York, where indeed the soils of the sandstone plateau and the severity of the climate make grass a more suitable crop than grain or fruit.

Butter making, like many other forms of manufacture, has its by-products, such as sugar, albumen, and casein. The last is used for food purposes, as a basis for paint, or is solidified for buttons, comb handles, electrical insulators, and for various other uses.

43. Cheese. The people of the United States consume more butter per capita than any other nation. They are, however, small users of cheese as compared with most peoples of Europe,

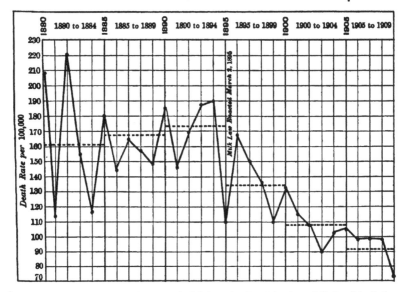

FIG. 36. Decrease in death rate of children in the District of Columbia following the enactment of the milk law of 1895. Dotted lines show averages for five-year periods (four years from 1905 to 1909)

and the making of cheese has been elaborated in that continent to a degree unknown in this country. A recent bulletin of the Department of Agriculture gives descriptions of two hundred forty-two varieties of cheese, made in various countries, chiefly in Europe. On the other hand, before 1900 the United States Census made no attempt to classify the brands of domestic cheese, most of which was known simply as American cheese.

As with butter, the making of cheese began to be organized about fifty years ago, and the first factory was built in 1851 in

Oneida County, New York. Canada now greatly outranks the United States as a producer and exporter. In fact, there has been a serious decline in exports of American cheese, believed to be largely due to loss of confidence abroad, by reason of skim-milk products and cheeses in which other oils were substituted for butter fat. In late years also America has faced strong competition in cheese brought from Australia and New Zealand to Europe. New York, Wisconsin, Ohio, Michigan, and Pennsylvania are the chief producers, and various brands of fancy cheese are now made to a considerable extent.

44. Dairy products of the United States at the Paris Exposition of 1900. The United States was the only country which attempted a continuous exhibition of fresh dairy products during that season. This was made possible by having a large display refrigerator, and also by a system of portable refrigerators packed in New York, shipped in cold compartments to Southampton, and thence via Havre to Paris. No other country provided such facilities. Milk was kept by no other means than cleanliness and cold, and was sweet from fourteen to twenty days after being drawn. Much skepticism was created until convincing proofs were given, for all French milk was soured by the second or third day, and the American samples had come from three thousand to four thousand miles.

Russia and Denmark were conspicuous, with the United States, in exhibits of butter; the great cheese displays were by Holland, France, Switzerland, and Italy; while the United States gave proof of its inventiveness in industry by showing a series of by-products and by offering representations of creameries and cheese factories in Iowa and New York.

45. General view. In both production and distribution comparison may be made between wheat on the one hand, and beef and dairy products on the other. In both cases the most of the world's international supply is produced on the great plains of temperate latitudes, though cattle often extend far beyond the margin of wheat into land too dry for that crop. So also the industrial nations of Europe, such as the United Kingdom and

Germany, with large populations and small territory, offer the best markets. Northwestern or Teutonic Europe produces much more of its own meat, butter, and cheese than of its own bread. Denmark, for example, imports its wheat, but sends out large exports of dairy produce. Norway, in like manner, buys its breadstuff but has its own herds and dairies. Cheap transportation is equally important in all cases, but swift transportation is more essential with meat, butter, and milk, though even here the invention of cooling appliances increases the allowance of time.

CHAPTER IV

IRON

46. Iron a representative mineral product. Wheat and cotton are produced directly from the soil by the aid of the processes of plant life. Cattle arise from the soil, but must have the substances of plants for their food. These and many others are living or organic products. We now take an example of the mineral substances which are useful to man, and which form much of the material of commerce.

About one twentieth of the earth's crust is iron. If it were all available, man could use but a small part of it. Most of the iron, however, is finely scattered among the rocks and soils, often giving them their color, or it is found in ores of such low value as to forbid profitable working. Iron seldom occurs in a free or metallic state, but is usually in composition with other minerals, forming an " ore," which is a raw material and must pass through processes of manufacture. The ore occurs in veins, sheets, or irregular masses at various depths below the surface, and is generally won by underground but sometimes by surface operations.

Special works give detailed information about iron ores. It is enough here to name the principal kinds, such as the red hematites, which are the most abundant; the brown hematites, otherwise known as limonites and bog ores; magnetite, an ore which is magnetic and contains a high percentage of the metal; and the carbonates.

47. Uses of iron. These are well known and need not be given in detail. We notice only certain great classes of uses to which this most important of all mineral substances is put. We include tools and utensils for almost all domestic uses and handicrafts; nearly all weapons used in warfare and in hunting;

building materials, from the frames of modern structures to the "hardware" used abundantly in all buildings. Further, iron is fundamental in modern transportation, whether we consider tracks, bridges, locomotives, cars, ships, wires and cables, or all kinds of road vehicles, such as carriages, wagons, motors, and bicycles. As the iron thus used is now in the form of steel, it is tough and strong, and the size of engines and cars can be increased several fold. These changes mean increased loads, greater speed, and cheaper rates, with a larger range of objects available to the user. Iron is a central theme in the geography of commerce. Modern machinery in its wonderful variety and effectiveness is made of iron. We have machines of iron for developing power, such as steam engines ; for transmitting power, as the wires and towers used in conveying electrical energy ; and for applying power, such as looms, lathes, printing presses, and the appliances of rolling mills.

48. Iron a measure of civilization and progress. The chief things belonging to progress are done with the aid of this metal. This is true not only of material advancement but also of intellectual and moral progress, through our present diffusion of knowledge and easy communication among individuals and nations. It has been truly said that the standing of an industrial people is exactly registered by the amount of iron it uses, and it is true that the present is the "iron age." According to a recent writer the steel industry is the "most important and accurate of all gages of the position of a people in the scale of civilization." The making and consumption of iron for each individual in the United States amounted in 1820 to about 40 pounds. This rose to 175 pounds in 1870, and to nearly or quite 400 pounds per capita in 1900. Some European nations even, to say nothing of more remote peoples, use less than 50 pounds of iron per capita.

49. Three periods of the making of iron in the United States. Charcoal was the chief fuel used in iron reduction and manufacture down to about 1850. This was possible because forests were then abundant, and the wood thus used must otherwise have been burned to remove it. The present high prices of timber are prohibitive of this process except in a few localities. In the second period, from 1840 to 1860, anthracite coal was largely used. This led to the first large production of iron in America. As the anthracite coal was confined to eastern Pennsylvania, so the iron interest centered there, — in the Lehigh, Schuylkill, and Susquehanna valleys, — and Philadelphia was the central

FIG. 37. Furnaces near Pittsburg

market. But this coal was an "obdurate" fuel, and has now lost its place in iron making, partly also because it commands a high price for household uses.

The era of bituminous coal and coke came in strongly about 1875, and extends to the present time. This change of fuel caused a shifting of the iron center from eastern to western Pennsylvania, in and around Pittsburg, where the soft coal is abundant and cheap. For the same reason the industry has stretched out along the whole course of the Appalachians, especially as both iron ores and bituminous coals are available at many points.

50. Deposits of iron ore in the United States. New York in 1889 mined 1,247,000 tons of ore, but in 1902 little more than 500,000 tons. This was due not to the working out of deposits but to the competition of Lake Superior ores, which are more easily and cheaply secured. Iron is mined in western and central New York, but more largely in the Adirondack region, particularly about Port Henry on Lake Champlain. Pennsylvania in 1880 produced more ore than any other state, but dropped to sixth place in 1902. It must not be forgotten, however, that while losing rank in iron mining this state keeps its primacy in iron making, having produced nearly one half the pig iron made in the United States in 1905. The Cornwall deposit in Lebanon County, a magnetite, is the chief center for the ore in Pennsylvania.

Virginia and West Virginia produced nearly 1,000,000 tons of ore in 1902, and Tennessee in the same year mined 874,000 tons. Alabama is the chief iron producer in the Appalachian belt, both for mining and manufacturing. The center of the industry is Birmingham. Most of the ores are red hematites, and 3,574,000 tons were mined in Alabama in 1902.

The largest production of ore in the United States is in the Lake Superior region, chiefly in Minnesota and the northern peninsula of Michigan. These two states in 1902 produced more than three fourths of all the iron ore mined in the United States, while in that year the single state of Minnesota mined more iron than the entire United States in 1889.

The five great "ranges" or ore belts in the Lake Superior region are as follows: the Marquette and Menominee ranges in northern Michigan; the Gogebic range, reaching from Michigan across the border into Wisconsin; the Mesabi and Vermilion ranges in northern Minnesota. The Michigan ranges have been worked for half a century, the first shipment of ore having been made, it is believed, in 1856. The Vermilion range has yielded ore since 1884, and the greatest producer of

all, the Mesabi, was not opened by a railway until 1892. In
1902 the five ranges had produced 221,000,000 tons of ore,
of which 54,000,000 came from the Mesabi range in the short
period of ten years. No foreign country has produced as much
ore in one year as these five ranges about Lake Superior.

The iron here occurs in large bodies, which in some cases are
so near the surface that the ore is mined in open pits. Such

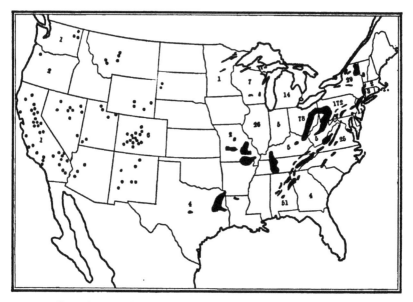

FIG. 38. Distribution of iron deposits in the United States

Minor areas in the east are omitted; Cordilleran localities are shown by conven-
tional signs. The size of areas does not indicate importance or the amount of
production. The numerals indicate the number of stacks of blast furnaces in the
various states. (After Harder)

excavations in the Mesabi range are so large that a network of
railway tracks reaches to the ore banks, and the ore, loosened
by a moderate amount of blasting, is loaded upon the cars by
powerful steam shovels. Not above two or three minutes is the
time needed to load a single car, and this fact greatly cheapens
the production of the ore. In fact, in the Superior district in
1902, 1156 tons of ore were won for each wage earner employed.
The cost of mining on the Mesabi range is from twenty to thirty

cents per ton, and this single group of mines may be considered as largely responsible for our great advance in the world's steel trade.

There are considerable deposits of iron in many other states, and they may become of more importance as the greater ore bodies of the Lake Superior region are worked out. Ohio and Kentucky are among these states, also Missouri, Arkansas, and Texas. Most of the Cordilleran states also contain iron, in deposits which are but partially developed on account of the newness of the region.

51. Transportation of iron. In the industries already noted we have seen the importance of transportation, and it is a no less significant element in the study of iron. Much ore in the Superior region is lifted from its native position by steam shovels and deposited in railway cars. It is then transferred to docks on Lake Superior and dumped into bins. These bins in turn discharge the ore by gravity into the holds of large lake steamships. One of these ships may take in tow two barges, also laden with ore, so that a minimum expenditure of power carries the cargoes down the Lakes, chiefly to Lake Erie ports. There it is lifted, often by "automatic unloaders," and much of it is transferred to cars and shipped to the furnaces of Pittsburg. As it is then handled by mechanical means in its passage through the blast furnace and the steel mill, it will be seen that the ore may not be lifted by a human hand in its entire course from its original resting place until it comes out a finished piece of steel, ready to be laid down in a railway track or raised into the framework of a building.

Transportation is, of course, not complete until the manufactured product has been distributed to the purchaser and user. This often involves carriage to a seaport, and thence by ocean steamship to foreign markets. It involves also the movement of pig iron to many mills for the making of machinery, agricultural implements, and all forms of builders' hardware, and these processes are in turn followed by another series of movements to myriads of final destinations.

FIG. 39. Duluth, Mesaba, and Northern ore docks, Lake Superior

The distance from many of the Superior mines to Pittsburg is about one thousand miles. The carriage is cheap because of effective transfer at the ports and the immense traffic on railway lines especially designed and exclusively used for ore. It is also to be observed that the long haul of the ore permits a short haul of the fuel, and that Pittsburg, Cleveland, and Buffalo are on the way to the great markets of the East and of

FIG. 40. Ore unloader at work, with bucket open. Lackawanna Steel Company's plant, Buffalo

foreign lands. The Lake ports from which the ore is shipped are Marquette, Ashland, Duluth, and Two Harbors on Lake Superior, and Escanaba on Lake Michigan. Among receiving ports are Chicago and Gary on Lake Michigan, and on Lake Erie, Toledo, Huron, Lorain, Cleveland, Ashtabula, Conneaut, Erie, and Buffalo. Return cars from Pittsburg do not necessarily run empty, but carry coal to the Lake ports, and may even bring back finished steel for shipment down the St. Lawrence and to Europe.

It should be further observed that lake transportation is cheaper than ocean carriage because the needed coal bunkers are small and the cargoes correspondingly greater. Contracts have been made at fifty cents per ton for transfer from Lake Superior to Lake Erie ports. Thus a ton of ore may be carried twenty miles for a single cent, a cost which is almost nominal.

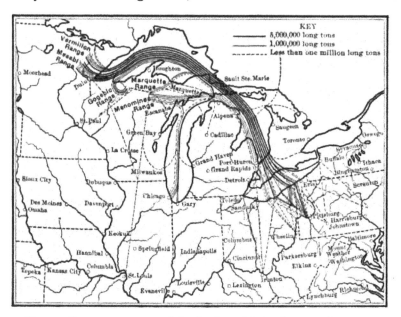

FIG. 41. Sources, routes, and destination of Lake Superior iron ores

Lines show amounts in long tons for one year. Lake Erie ports which bear no name on the map are as follows, from Erie westward: Conneaut, Ashtabula, Fairport, Cleveland, Lorain, Huron

In the Appalachian fields the ore, the fuel, and the limestone used as flux are all near together. The larger item of transportation is in the marketing of the product, much of which must be carried to the Northern and Western states.

52. American centers of smelting and steel making. Pittsburg is the chief of these, and the general conditions which have favored its development have been already set forth. By Pittsburg we mean the region having this city as its center and extending out from fifty to one hundred miles and even embracing the adjoining portions of the state of Ohio. Of every five tons of iron ore mined about Lake Superior, three come to this region. Allegheny County, to which Pittsburg belongs, produces one fourth of the pig iron of the entire country and even a larger part of our steel.

Most of the Lake ports which receive ore also smelt it and make steel. South Chicago and Joliet, in Illinois, afford an example of long transfer both of ore and coke, the former coming from Lake Superior and the latter from Pennsylvania. There is, however, the compensating advantage that the product is near to the markets of the Mississippi Valley and the upper Lake region.

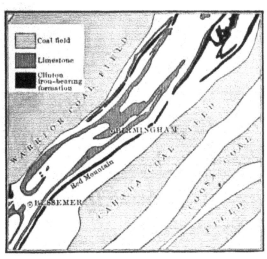

FIG. 42. Birmingham district, Alabama, showing relation of iron ore, limestone, and coal

In the southern Appalachian region Birmingham is the iron center. Among the products are steel rails, rods, nails, all kinds of wire fencing, plows, cast-iron pipe, sugar-house machinery, bridges, engines, boilers, and stoves. In lesser degree, but with similar advantages, the iron industry has established itself in Chattanooga and Knoxville in eastern Tennessee. It should be observed that many iron centers in Alabama, Tennessee, Virginia, and Pennsylvania lie in the Appalachian valley.

53. Processes and products in the iron industry. The essential part of a blast furnace is a lofty structure which receives " charges " of ore, coke, and limestone at the top, and from which flows, at the bottom, the molten iron and the waste or slag. Iron ore may contain 40 to 70 per cent of metal, and this must be separated from the other minerals combined with it. For this purpose great heat is needed, and limestone, which serves as flux, combining with the nonmetallic materials and flowing out with them. The air fed to the furnace is first raised to a high temperature, the furnace itself being lined with fire brick. At certain intervals the iron is drawn off and may be run into molds, forming " pigs." In modern processes, how-ever, most iron is carried in a molten state to the " converter," where it is turned into steel. Even thus, however, it is common to call it pig iron, as if it had been molded. A great saving is effected because the iron need not be remelted for the making of steel.

FIG. 43. Ore boat below the " Soo," returning empty

Pig iron is brittle, being neither ductile (capable of being drawn) nor malleable (capable of being hammered into various shapes). Cast iron is brittle and is used for some purposes, as stoves. Wrought iron is so treated as to contain little car-bon, and is both malleable and ductile. Steel contains more carbon than wrought iron and less carbon than cast iron. It can be made into any desired form when hot, and when cold is in-tensely hard, flexible, and strong. It has taken the place of other forms of iron for most uses, and is made in a variety of alloys, according to the purposes for which it is intended. Thus a soft steel would be used for making nails, and a steel of " invincible " temper for making tools.

In early days steel was made by surrounding iron with fine charcoal and heating it for several days. This was an expen-sive process, and the product could not be used for many pur-poses. Sheffield steel sold for $250 per ton, and at that center fifty years ago 50,000 tons were made in a year. At the present time many million tons of iron are annually made into steel, and the product may be sold as low as $25 per ton. This

revolution became possible through the invention of the Bessemer process, by Sir Henry Bessemer, an Englishman. The molten iron goes directly to a large receptacle called the converter.
There it is still more intensely heated, and " spiegel " is added, —
a compound containing the proper amounts of carbon and man-
ganese. The newly made steel is molded into large blocks or
ingots, and then, still intensely hot, may be sent at once to a
rolling mill, to come out as steel rails. Thus there is no break

FIG. 44. General view of plant, Lackawanna Steel Company. Ore docks and
blast furnaces are on the right

in the process, and there is 'no cooling between the conditions
of crude ore and finished steel. A more recent method, now
extensively used, is known as the open-hearth process.

Ores are often termed Bessemer and non-Bessemer, according as they are or are not suitable for steel. Any but the small-
est percentages of sulphur or phosphorus are not allowable in
the production of steel. The modern processes have led to an
enormous growth of iron industries in all advanced countries,
and in particular have given to the United States the leadership
in this field. " American primacy " has been ascribed to three
special advantages : first, the great supply of ores and of coking

coal; second, superior industrial organization; and third, a large home market, which gives uniform conditions to the trade, relieving it from close dependence on the tariffs of foreign countries. Under the head of organization is to be counted the

FIG. 45. Bessemer converter in action, Lackawanna Steel Company

consolidation of the various departments of the industry under single great corporations. Thus a single corporation may own or lease mines in Minnesota, and own the Lake steamships, the railways at either end of the water journey, and the furnaces in western Pennsylvania.

54. Reserves of iron ore in the United States. At a conference of governors and experts held at the White House in May, 1908, the conservation of mineral deposits was discussed by Mr. Andrew Carnegie. Several statements are here taken from his address. In 1907 there were mined in the United States alone 53,000,000 tons of iron ore, or more than 1200 pounds for every person included in our population. It is

FIG. 46. View on pouring side of open-hearth mill, Lackawanna Steel Company

estimated that 1,500,000,000 tons of ore then remained in the Lake Superior district, and 2,500,000,000 in the Southern district, including Alabama, Tennessee, Georgia, and Virginia. The estimate for the rest of the country was from 5,000,000,000 to 7,000,000,000, giving a total reserve supply of about 10,000,000,000 tons. It was shown that at the present rate of use the Superior supply would be exhausted before 1940, or in one generation. About one thirteenth of the estimated original supply has been mined. With probable increase of use, the present supply of workable ores will be gone long before the end of the present century.

An example of such exhaustion is found in the case of Iron Mountain in Missouri, whose ores were once thought to afford a permanent supply, but have now been worked out. Mr. Carnegie urges that the demand on iron must be lessened, particularly by substituting water carriage, where possible, for railways. Railways, with their tracks and rolling stock, call for much more iron than steamships, for transporting the same amount of freight. Buildings and bridges also may be made of concrete rather than of iron, and the raw materials of the former are inexhaustible. Other metals and new alloys of iron may also be used.

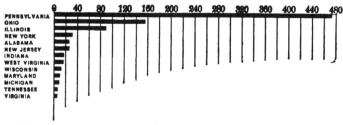

FIG. 47. Value of iron and steel products of leading states in 1905 in millions of dollars. The figures for West Virginia, Maryland, and Tennessee are for 1900

Mineral deposits are fixed in quantity. If used up, no more can be produced. Thus they differ from the materials of vegetable or animal industries, in which, with reasonable care to preserve and manage the soil, production is unlimited and may serve the needs of the human race for an indefinite time to come. With minerals the supply is fixed, and we must depend on economy, on finding new deposits now unknown, and on the invention of new applications and combinations.

55. Foreign iron trade of the United States. The iron products exported from the United States increased in value from $14,000,000 in 1880 to $181,000,000 in 1907. An example is found in steel nails, of which we send out 25,000,000 pounds annually. The change in price from Bessemer's day (10 cents) to ours (2 cents) illustrates the cheapening of many useful things through the development of work in iron. American typewriters are found in every nation of the world. Locomotives are another important item in our foreign trade in steel, as are bridges and machinery of all kinds. Agricultural machinery was sent out to the value of nearly $27,000,000 in 1907, marking a fivefold increase from 1897.

Our growth in steel making not only permits of exports of steel products, but is at the foundation of general success in manufactures. We make more than one third of the manufactured products of all nations, and thus, through the medium of effective machinery and transportation, all our exports are largely increased by American success in iron. In 1870 Great Britain made four times as much pig iron as the United States, but our country passed Great Britain as a producer of iron in 1890. Even in that year our exports were but $28,000,000 and

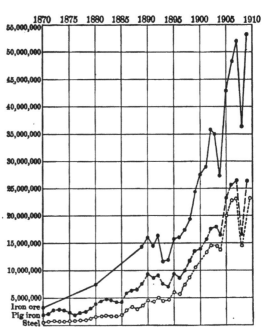

FIG. 48. Production of iron ore, pig iron, and steel in the United States, 1870–1909, in long tons

our imports were $53,000,000. In 1900, however, the ratio was very different, and we exported iron products to the value of $121,000,000 and imported them to the value of $20,000,000. Thus in iron we have become nearly self-sustaining, and are supplying a growing surplus to other nations.

56. Iron in Great Britain. In 1906 there were produced in the United Kingdom 10,109,000 tons of pig iron. The United States in the same year produced 25,307,000 tons. The iron ore mined in Great Britain in 1906 was 15,500,000 tons, and 7,823,000 tons were imported. Of the imported ore much comes from the northwestern or Biscayan provinces of Spain, and in later years importations from Sweden have been on the increase. The British iron industry is able to rely so largely on imported raw material because the British coal supply has thus far been ample ; and this, added to British skill and to extensive plants already in operation, has offered reasonable growth even in the face of the dominance of the United States and the growing competition of Germany. Many of the British iron centers are on the coal fields, and in no case is the distance from coal so great as to be a serious handicap where transportation both by land and water is so highly developed as in the United Kingdom.

As in the United States, the early fuel was charcoal, and hence the blast furnaces were, as stated by Mackinder, in the Weald of Kent and Sussex in the southeast, in the forest of Dean in the west of England, not far from Gloucester, and near the present great center of iron, Birmingham, in the forest of Arden. According to the same author coal was first used largely in smelting, in the eighteenth century. As coal was found near Birmingham, it naturally took the place of charcoal as the forests failed, and thus iron making continued in the same locality. Birmingham and Wolverhampton are centers of what is appropriately called the " Black Country," and turn out manufactured objects of iron in great variety. At Sheffield also the old iron industry grew into the new, and was favored by the presence of water power and even in larger degree by water carriage of Swedish iron,[1] which was suitable for making steel by the older methods, which in earlier days prevailed in that center for cutlery. Sheffield has now added work in armor plate and steel rails, and is unrivaled in making " high-speed " steel, — that is, steel used in tools which are operated with great rapidity and therefore must suffer heavy strain. The cause of Sheffield's superiority in steel of higher grades probably lies in

[1] Mackinder, Britain and British Seas, p. 325.

accumulated experience and in the thorough care given to each process of production.

The iron industry has flourished on a large scale about the lower courses of the Tyne, the Wear, and the Tees. Near the mouth of the last is Middlesbrough. This iron town is on the northern edge of Yorkshire, was founded in 1830, and is near the ore beds of the Cleveland district. English capital built a town of the same name in southeastern Kentucky, at the Cumberland Gap, in the hope that it would become a great center for iron.

Newcastle and a chain of settlements along the estuary of the Tyne, close to extensive beds of coal, rival the Clyde near Glasgow in the vast shipbuilding operations of Great Britain, — and shipbuilding in modern days is chiefly a business in steel. The Clyde offers the greater shipbuilding industry not only of Great Britain but of all nations. Its artificial deepening has enabled Glasgow, the second city of the United Kingdom, to grow up upon it, and its iron and other industries have been favored by the presence of iron and coal in the neighboring central lowland of Scotland. The iron industry is strongly developed in the towns of the South Wales coal field. Like the districts of the Clyde and the east coast, ocean transportation is available, while Birmingham, an interior town, is the center of a network of railways.

57. Iron in Germany. The German Empire is the first industrial nation on the continent of Europe, and the greatest rival of the United States and Great Britain in the field of manufacture. The largest item in German manufacture is iron, and here, as in other iron-making countries, there is a close relation among the three factors, iron ore, coal, and transportation. By building canals and improving her rivers, Germany has made the most of her internal navigation. Her natural advantages also have been used with energy and skill in a land where education of every kind, including technical training, has been brought to a high degree of perfection. The lower Rhine country, in western Prussia, is the greatest single seat of the iron industry. A short distance north of Cologne, a small river, the Ruhr, enters the Rhine from the east. In its valley is a coal field, one of the greatest deposits of this mineral in Europe. Here also is Essen, best known as the home of the Krupp steel industries, founded and developed by men of this name during three generations. The works were originally begun to compete with the steel of Sheffield. Plants at several other places now belong to the same establishment, which has many mines in Germany, some in Spain, maintains wharves at Rotterdam, and owns railways, telegraphs, and proving grounds for its famous guns. It is thus like some of the vast corporations of the United States in its comprehensive ownership of all branches of the industry, and its motto is said to be, "The establishment itself must furnish whatever it requires." Among the more convenient sources of its ore are the Harz Mountains, not far to the southeast, and Alsace-Lorraine and the grand duchy of Luxemburg, from which the Rhine and its western tributary, the Moselle, offer a natural highway to the Essen district.

Shipbuilding is one of the largest items in German iron manufacture, and the rivalry between Germany and Great Britain in this field is well known. Among other seats of this industry is Stettin, in northern Prussia, on the Oder, and not far from the Baltic. The Oder is navigable, and through the aid of canals offers access to the basin of the Elbe and to much of northern Germany. We may compare this center with Essen near the Rhine, Newcastle upon Tyne, or Glasgow on the Clyde. Iron ores are found in several other parts of Germany, although in no case are they so important as those of the western border.

58. The iron industry of other nations. Several other European nations have important iron industries, depending in part on population and needs, and in part on their resources of iron ore and coal. Thus France produced in 1906 over 8,000,000 tons of iron ore. Her largest home supply is from the basin of the Moselle on the east, a region adjacent to great German deposits and to those of the grand duchy of Luxemburg ; nor is it far from the iron and coal which have made southern Belgium so important in industrial pursuits. From these bordering districts, and from Spain, France supplements her home supply.

Spain mined in the year 1905 over 9,000,000 metric tons of iron ore, much of which was exported to Great Britain, Germany, France, and other countries While the annals of mining in Spain go back to the Phœnicians of ancient days, she has not attained the industrial strength of nations like Belgium and France, which are less favored in mineral resources. Spain therefore shows that the spirit and genius of the people are of equal importance with natural resources ; otherwise she would not let her coal remain unused in its beds, and send out her iron as raw material to be smelted and converted into steel by her more progressive neighbors.

The United States appears to be the one country in the world which at the same time has iron resources of the first order and is utilizing them to the highest degree. Great Britain, Germany, France, and Belgium exhibit skill and enterprise, are working their own deposits as fully as possible, and make up their home deficiencies by importation. Spain has a surplus because of her indifference. Italy is an example of countries having, as far as is known, insignificant supplies of ore ; and she must therefore, with her large population and her industrial awakening, import nearly her entire supply of iron and the products of iron manufacture. Sweden appears to be the one nation of Europe which has ample iron, is fully awake to the importance of its home manufacture, and at the same time is able to supply a surplus to the nations of western and central Europe. The abundant forests of that country enable the charcoal process to survive more fully there than in most other regions. The iron industry is growing in Russia and Austria-Hungary, and these countries have now become producers of the second order. Great nations like Russia or India may have extensive agriculture before gaining the skill and organization essential to work in metals.

Three countries — the United States, Germany, and Great Britain — are the chief competitors in the iron trade, and are likely so to remain for a considerable period in the future. This view is warranted by their resources, their skill, and by the hold already gained upon the market.

The chief advantages of England are the proximity of ore, coal, and flux, and the nearness of the iron centers to the sea, which makes them accessible to all markets and makes it possible to import ore at small cost. A home authority has recently expressed full confidence in British iron, urging that the coal supply is ample and that Spanish ore can be transported at half the cost of carrying Lake Superior ores to Pittsburg. Should the Spanish supplies be exhausted, ample reserves are found in Scandinavia, Lapland, Algeria, and many British colonies.

Much of the German ore is of low grade, but is well used by modern methods. Germany has the largest coal fields of Europe, but the coal is not so close to the iron as in England, where, according to a German authority, about 10 per cent of the cost of pig iron is to be charged to transportation of raw materials, as against 30 per cent in Germany.

The United States has much larger reserves both of iron and coal, but the distances of transportation both for raw materials and finished products are usually greater. Ore, coal, and flux are close at hand in the southern Appalachian district, but the greater markets are remote. The development of the southern states, however, will open a large market, and trade may develop in Latin America, especially through the opening of the Panama Canal.

CHAPTER V

COAL

59. Coal in Pennsylvania. This state is first in the production of iron and steel, though not of iron ore. Its leadership in iron is, however, due to its primacy in coal, and hence we begin here our study of this most important of the fuels of the world. Pennsylvania affords by far the largest mineral output of any state in the Union, having produced from mines and quarries in 1902 values amounting to $236,000,000, while the second in rank, Ohio, produced values of $57,000,000. The coal product of Pennsylvania in 1901 was about 150,000,000 tons.

Eastern Pennsylvania contains the anthracite region of the United States. Outside of this state the hard coal is found in small amounts in Colorado and New Mexico, but this western product is not commercially important. Western Pennsylvania contains the northeastern parts of the bituminous fields of the Appalachians and the Mississippi Valley, and it is here that Pittsburg, the first coal and iron city of the world, has arisen. Bounding the anthracite field by well-known features, it is contained in a rude quadrangle having the Delaware and Susquehanna rivers on the east and west, the north branch of the Susquehanna on the north, and the long ridge of Blue Mountain on the south. There is short and down-grade passage to New York, Philadelphia, and Baltimore, and moderate distances include New York, Massachusetts, Connecticut, Rhode Island, and New Jersey, the five states of greatest density of population. Here also, partly because of accessible coal and partly for other reasons, are regions of the highest industrial activity.

The anthracite lies in several geological troughs or downfolds of strata, often designated as the northern, middle, and southern basins. Important groups of towns have grown up in response

79

to the mining industry. In the northern basin are Scranton, Carbondale, Pittston, Wilkes-Barre, and Nanticoke. In the middle basin are Hazleton and Shamokin, while the main center of the southern field is Pottsville. The area actually occupied by workable coal measures is 484 square miles. The anthracite output of 1901 was 67,000,000 tons.

So far as is known the Pennsylvania anthracite was first used in 1769 by a blacksmith whose forge was near the place now

FIG. 49. Truesdale coal breaker, Scranton, Pennsylvania

occupied by Wilkes-Barre. From 1776 to 1780, in the time of the Revolution, anthracite was shipped to Carlisle, in southern Pennsylvania, for the manufacture of materials of war. In the early years of the nineteenth century this coal found its way in " ark " boats and wagons to Philadelphia, being used for graveling walks, and then in grates and furnaces, but it is on record that most of one shipment was given away. In 1823 the first cargo of anthracite was sent around Cape Cod to Boston. These historical facts show how recently coal began to have importance as an article of commerce.

About 15,000 square miles in western Pennsylvania are more or less underlain by beds of bituminous coal, which lies along the Monongahela and Ohio rivers, continuous with deposits in Ohio, West Virginia, and Maryland. The first record of mining goes back to 1760, and the place was across the Monongahela River from the site of Pittsburg. The advent of steam soon caused a demand for coal, and by 1800 it was used to some extent in local factories. Coal was first shipped from Pittsburg in 1803. In 1902, or ninety-nine years later, there were more than 1000 mines in the bituminous beds of Pennsylvania, and the output was nearly 100,000,000 tons, — more than one third of the bituminous coal production of the United States. The anthracite coal belongs to a region of disturbed rocks in the mountainous part of the state, while the bituminous beds and the rock strata associated with them lie in their original horizontal position. Thus geological disturbances have given different qualities to coals of the same age, and have also led to differences in the methods of mining in the two regions.

60. The market for Pennsylvania coal. The coal provides power at the pit's mouth for elevators and drills and the machinery of the "breakers," by which the product is reduced to assorted sizes. It cooks the food of the miner and warms his home. It fires the locomotives which haul the fuel to Philadelphia, to many towns of the state, and to innumerable points beyond the boundaries. It feeds the furnaces of Pittsburg and of all the lesser centers of iron and steel from the Delaware River to Lake Erie. It furnishes the chief freight for the Pennsylvania, the Lehigh Valley, and other railway systems.

The state of New York, with more than 9,000,000 people, and the city of New York are close at hand. The homes, factories, and railways of the Empire State must have coal, and there is none within its boundaries. The Delaware and Hudson, the Lackawanna, the Ontario and Western, the Lehigh, and the Pennsylvania railways all run from the heart of the coal regions far into New York, to Lake Champlain, to Utica and Watertown, to Syracuse and Oswego, to Rochester and Buffalo.

Piers on Lake Ontario and Lake Erie receive Pennsylvania coal, and transmit to much of southern Canada its chief supply of fuel. Within the United States, Pennsylvania coal is hauled as far west as the Missouri River and beyond the Great Lakes, both because the Mississippi Valley has no anthracite and because Pennsylvania soft coals are suitable for making coke. Transported to New York City the coal brings heat and power for its

FIG. 50. Coal-storage plant of the Delaware and Hudson Company, Delanson, New York

vast interests and fills the bunkers of steamships departing daily for all parts of the world.

61. The Appalachian coal field. Pennsylvania coals lie in the northern parts of the Appalachian Mountains and the Appalachian plateau, and thus belong to a succession of deposits which extend from eastern Pennsylvania to central Alabama. This field is about 850 miles long, lies in parts of nine states, and includes about 70,000 square miles. It covers a wide belt in eastern Ohio, West Virginia, and eastern Kentucky, with smaller but important areas in Maryland and Virginia. Its chief southern

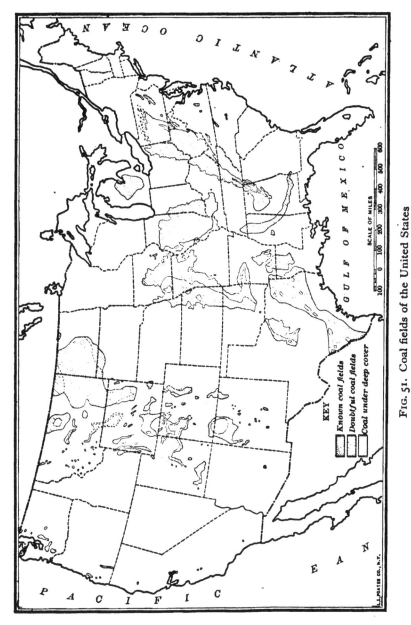

FIG. 51. Coal fields of the United States

KEY

Known coal fields
Doubtful coal fields
Coal under deep cover

parts are in Tennessee and Alabama, with a slight development in the northwest corner of Georgia.

The coal from this field has a wide market, as has been seen in part for the coal of Pennsylvania. West Virginia and Ohio coal reaches markets on the Ohio River, the Great Lakes, and along the lower Mississippi River. Several railways convey coal to the Atlantic seaboard, to ports extending from New York to Savannah. Alabama and Tennessee fields are well placed to supply the Gulf States and to export to Central and South America.

62. Interior coal fields. The northern interior field occupies about 7000 square miles in the central part of the southern peninsula of Michigan. This coal goes mainly to local markets. It cannot move south and east, being unable to compete with the better grades of Ohio and Pennsylvania. It does to some extent cross Lake Michigan to Wisconsin, and even to states beyond the Mississippi River, since westward freights are cheaper than eastern, owing to the greater amount of bulky commodities moving toward the east.

The eastern interior field is largely in Illinois, extending into southwestern Indiana and Kentucky, and includes 46,000 square miles. Its markets are chiefly local, being on the field or in adjacent territory. Naturally this field largely supplies Chicago and St. Louis, the most of Indiana, western Kentucky and Tennessee, and some territory west of the Mississippi River. This coal competes with that from Pennsylvania. In the summer months soft coal comes by the Great Lakes from Pennsylvania to Chicago, and kills the Chicago trade from the home field, which revives again with the closing of navigation in the winter.

The western interior field includes parts of Iowa, Nebraska, Missouri, and Kansas, and covers more than 60,000 square miles, about two thirds of the area being considered productive. This field continues, under the name of southwestern, into Oklahoma, Arkansas, and Texas. Here again the market is largely local, but tends to extend westward until stopped by competition with the coal of the Rocky Mountain fields. The southern railways crossing the continent, and those of Texas, get their supplies from the southwestern fields. The general tendency of movement is toward the west, and this is ascribed [1] to three causes: first, the eastern coals are better than the western; second, freight rates are generally lower westward than eastward; and third, water transportation favors westward carriage. This is particularly true of the Ohio River.

63. Other coal fields in the United States. In Virginia and North Carolina are small areas of bituminous coal, known to the geologist as of Triassic age. The beds were opened and slightly worked in early days but are now unimportant. In the Gulf States, from Alabama to Texas, are belts of lignite, which the needs of the future may raise to commercial importance. All coal is modified from wood or other vegetable tissues. Anthracite is much changed, bituminous coal in less degree, while *lignite*, as the name implies, retains much of the original character of wood. Other coal fields (Cretaceous) are found along the Rocky Mountains, extending eastward on the Great Plains of Montana, the Dakotas, and Colorado, and south to the plateaus of Colorado, Utah, and New Mexico. The position of these deposits appears on the map (fig. 51). The important coal beds of the Pacific coast are found in the state of Washington, and there are small areas in Oregon and California.

64. Coal the foundation of modern industry. The great use of coal belongs to the past one hundred years. This does not mean that this fuel was not in use before, for its value was known in a small way in ancient times among the Greeks and the Chinese. It is known also that the early Britons and the Roman invaders of their land used coal, and by 1215 A.D. some trade had been established, making it an article of commerce. In the thirteenth century Newcastle coal was brought in ships to London, and soon the smoke of its burning aroused complaints because the air was defiled with smells and dust. In the seventeenth century the European peoples made more use of coal, and at the end of the eighteenth century the newly invented steam engine led the way to the world-wide developments of to-day, in which

---

[1] C. W. Hayes, *22d An. Rep. U. S. G. S.*, Part III, p. 24.

coal propels the machinery of most factories and supplies the chief power for conveyances on land and sea. It has been influential in locating the centers of manufacture and commerce, and Birmingham, Newcastle, and Pittsburg are conspicuous examples. Other causes have helped, however; for New England industry was located by water power, even though most of her machinery is now run by steam. New York could not maintain its industry without coal, though other factors than the Pennsylvania coal seams would have made New York America's metropolis. Coal makes transportation of raw materials and manufactured products easy, and hence tends to diffuse the centers of industry or allows them to remain at considerable distance from the coal, as is the case with Boston and many New England mill towns. The industrial importance of coal is nowhere better seen than in the case of Great Britain, where, to quote H. R. Mill, "It is coal which makes it possible to purchase grain and other food materials ; not directly, . . . but indirectly, by supplying smelting furnaces for reducing iron and providing power for engineering works and factories." The United Kingdom may thus continue to be an industrial nation, even though she imports iron, so long as her coal holds out. A recent British writer computes that, at the present output, the British reserves of coal would last for six hundred and thirty-three years.

65. The development of transportation dependent on coal. When the nineteenth century began, men of public spirit and enterprise were planning highways and canals. On the one were stages and freight wagons, and on the other were boats. Whether by road or canal, passengers and articles of commerce were moved by domestic animals. On the seas the only power was the fickle wind, which, mastered by the utmost skill of the sailor, required weeks for a voyage across the Atlantic and many months to sail around the world.

The first half of the century saw the perfecting of the steam engine. The Mississippi and the Hudson were covered with boats, and lines of railway led from the Atlantic seaboard across the Appalachians to the prairies. This was made possible by the use of coal. The second half of the nineteenth century saw a vast increase of steamships and of railways. The cutting of the forests made wood impossible as fuel, and thus this greater development depended still more closely on the beds of coal. For about seventy years steamship lines for regular service have been maintained across the Atlantic, and for forty years it has been possible to be moved from New York to San Francisco by the long-stored energy of the beds of coal. Swiftness, regularity,

Fig. 52. Coal barges at Pittsburg ready to move down the
Ohio River

and low freight rates are the three factors upon which the volume and wide distribution of modern commerce depend, and these have all been secured by modern inventions, which brought into employment a fuel long known but little used.

The sailings of battleships and of the merchant marine of all nations are largely conditioned by the coaling stations, which are now an essential part of the world's network of ocean traffic. Great Britain, in particular, maintains these depots in all latitudes and longitudes, — in the Atlantic, the Mediterranean, along the shores of India, Australia, and the islands and mainlands of the Pacific. A ship may fill its bunkers at Gibraltar or

Barcelona; at Genoa, Naples, Trieste, or the Piræus; at Port Said, Colombo, Singapore, Manila, Yokohoma, Honolulu, or San Francisco, and thus move step by step around the world. For present generations, at least, the world's transportation is mainly based on coal.

66. British coal fields and the centers of British industry. We take first the coal fields of England. These are six in number, as follows:

1. Newcastle, a field on the lower Tyne, and supplying this center; also the export trade of the neighboring Sunderland and Hartlepool, and the iron works of Middlesbrough.

2. West Riding, a field lying south, toward the center of England, supplying Sheffield, Leeds, Nottingham, and other manufacturing towns, and adjacent to the seaport of Hull. Both the fields just described send ·much coal to London, and both lie on the east of the

FIG. 53. Coal fields of Great Britain

Pennine chain, the great upland and backbone of northern England.

3. The Midland basin, which includes a number of small fields on the plains of central England, lying between London and Liverpool, and supplying Birmingham and the other industrial towns in its vicinity.

4. The Lancashire field, which is across the Pennines westward from the West Riding coal, supplies Manchester and other mill towns and provides much coal for the ships of Liverpool.

5. The Cumberland basin, west of the Pennines and opposite Newcastle field.

6. The southern coal fields, lying in the neighborhood of Bristol. Outside of England are two great fields :

1. South Wales. This field supplies the iron, tin-plate, and copper industries of Cardiff. From Cardiff vast quantities of South Wales coal are exported or sent to British coaling stations in all parts of the world.

2. The Scottish fields, in the Scottish lowland, adjacent to the firths of Clyde and Forth, supplying the industries of Glasgow, exporting to Ireland and the Baltic, and contributing to the greatest shipbuilding industry in the world.

67. Coal in other countries. Of all European countries except Great Britain, Germany has the largest coal beds and makes most use of them. Her fields are in the Ruhr basin, already noticed in the account of iron ; along the Saar, in the Moselle basin, near the French boundary ; in southern Saxony about Zwickau, not far from Leipzig and Dresden ; and in Silesia, near the Austrian and Russian border and extending into those lands. Breslau is the center to which these Silesian deposits are especially tributary.

France has coal in several areas, and one of these is in the northeast and continues into the great industrial region of coal and iron in southern Belgium. This coal area is close to the coal and iron of the whole Rhine region and of the grand duchy of Luxemburg, making this part of western Europe one of the first commercial and industrial centers.

Spain is well supplied with coal but awaits enterprise and development. Coal deposits are insignificant in Portugal, Italy, Switzerland, Holland, Denmark, and the states of the Danube and the Balkans. The important coal beds of the British colonies are found in Canada, particularly in Nova Scotia and other eastern provinces ; in South Africa, the greatest production being in the Transvaal ; in New South Wales and other provinces of Australia ; in New Zealand ; and in India, whose fields have been estimated at 35,000 square miles.

Coal beds are important and now much worked in Japan. Chinese coal fields are imperfectly known and little worked, but information warrants the belief that 200,000 square miles is a low estimate for their area. As coal is at the basis of industry, it will be seen how profoundly the commerce of Asia and of the world may be affected when modern enterprise has full opportunity in this great empire. Our survey of the world's coal is enough to show that this fuel is very generally distributed, for South America also has several fields, and exploration may

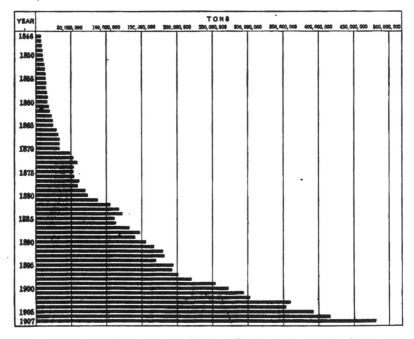

FIG. 54. Annual production of coal in the United States, 1846–1907

be expected to increase the number. This general supply of power will favor local prosperity and comprehensive systems of transportation and exchange, including all the habitable lands of the globe.

68. Varieties and by-products of coal. Coal is the remains of vegetation, and ranges from the soft peat of modern swamps to the hard anthracite of ancient forests. The soft or bituminous coal may be just as old as the anthracite, but has been subject to less of pressure or other influences that would change its condition. Lignite is intermediate between peat on the one hand and true coals on the other.

Special types of coal are sought for particular industries, and Pennsylvania coal and coke go far west in the Mississippi Valley, because suited to domestic needs and to some classes of manufacture. Thus, results of a distinctly geographic nature follow from the variety in the qualities of coal.

The number of by-products derived from coal is considerable. Among these is illuminating gas, and in some cases the manufacture of coke would not be profitable were not the gases utilized for this purpose. The by-products include also several sorts of chemical products, such as flavors, perfumes, dyes, soaps, antiseptics, cleaning preparations, and explosives.

69. The conservation of coal. Some of Mr. Carnegie's estimates at the White House Conference of 1908 are here given. The entire original supply of coal in the United States was put at 2,000,000,000,000 tons. We mined in 1907, 450,000,000 tons, or more than five tons for each person in the United States. At the present rate of increase the production in the year 1937 would be over 3,500,000,000 tons. Mr. Carnegie thinks that unless unforeseen economies or inventions intervene, most of our coal will be gone in two hundred years.

The methods of mining coal also cause much waste. Many beds of coal contain rock or earthy matter, which diminishes their value, and such coal is often thrown to refuse heaps in the mine or at its mouth. The average amount of this impure coal for all the mines of the United States is estimated at 25 per cent, and it is lost in spite of the fact that it is high in heating power and would for some purposes have great value. In some mining regions, by careless methods, 40 to 70 per cent of the coal is believed, on expert authority, to be left in the ground. According to Dr. I. C. White, the gas from 35,000 coke ovens in the vicinity of Pittsburg is wasted, which means a loss of one third of the power and one half the value of the beds of Pittsburg coal. Dr. White estimates that at the present rate of consumption this seam of coal will give out in 93 years. As the industrial greatness of Pittsburg depends on this coal, here is a fact of the utmost importance to the student of commercial geography.

Mr. Carnegie also shows that the methods of using the coal are more wasteful than those of mining, for in power plants only 5 to 10 per cent of the energy actually contained in the coal is used, while in electric lighting and heating only about one five-hundredth of the power belonging to the coal is utilized.

From these statements appears the importance of saving coal supplies in every way. The exhaustion of this fuel would introduce great change in the character and distribution of the world's industries. Only the progress of invention could in this event save the human race from limitation and hardship. The present generation owes the duty of reasonable thrift to those which will come after it, and the present waste is to be counted as little less than crime.

CHAPTER VI

THE PRINCIPLES OF COMMERCIAL GEOGRAPHY

70. Commercial geography defined. General geography comprehends the surface of the earth, with its life. Physical geography includes the forms of the land, the ocean, the atmosphere, and the globe in its movements and relations as a planet. Text-books of physical geography usually give some account of plants and animals, including man. If man himself is studied in his relation to the earth, the subject is termed anthropogeography. If within this field we, in a more special way, study man as a trader, and the earth's surface as affording a place to trade, the theme is commercial geography. It is clear that the student of this subject must deal with land forms, soils, climate, the distribution of water; with minerals, animals, and plants ; with routes, cities, and countries, — he must indeed bring all geography under tribute.

Thus commerce is largely geographical in its nature. It involves the distribution of all natural features and products. It is also, in its larger development, carried on among peoples, and is therefore largely controlled by historical, racial, and political conditions. Hence political and economic history must be taken into account, and, even prior to these, the discoveries which have gradually made the earth known and the migrations by which the lands have been peopled.

71. The development of commerce. The primitive community is sufficient to itself. It provides all its own food by hunting, fishing, or simple agriculture, and makes its own clothing of skins and rude fabrics. A dugout, a snow hut, or a tent of poles or grasses serves as a habitation, and the implements of domestic life and of warfare are of home production. Commerce begins when the primitive man finds in the possessions of another

tribe or race something that he can gain by barter or can win by
some medium of exchange which seems valuable to both parties.

FIG. 55. Distribution of population in the United States, 1790

The real development of commerce follows upon growth in
civilization and the multiplication of needs and desires. Then
man's life ceases to be a reflection of the things immediately
around him, and takes in a larger world. Along with rise in
the scale and complexity of living goes increasing knowledge

of other regions, travel, and evolution of means of transportation. Gradually experience teaches the economy of a division

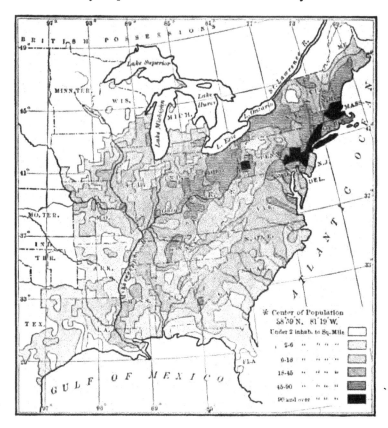

FIG. 56. Distribution of population in the United States, 1850

of labor, and the wisdom of each family, each tribe, or each
nation doing what it can do best, and producing what it can
produce to the highest advantage.

Commercial geography for an American savage would have
included a tract of forest or prairie and a few neighboring tribes.
His trade could hardly be called commerce until the white man
came, enlarged his knowledge, excited his desires, and began
to trade with him, receiving his peltry for clothes, knives, guns,
and trinkets. Commercial geography for the Roman people

would have centered in the Mediterranean region, reaching out to the Orient, to North Africa, to Germany, and to Britain. Commercial geography in the seventeenth century belonged to a few nations of Europe, exchanging fabrics and furniture for American tobacco, and continuing with the Far East, for silks and spices, the trade of medieval days. Now all peoples except a few remote tribes, all seas except those bound in polar ice, have to do with the production, transportation, and sale of commodities for those, near or far, who do not produce them.

It is now time to analyze more in detail the principles of the subject. These have in some manner been stated and exemplified in the study of the five typical products which form the themes of the preceding chapters. A fourfold division is made, and sections 72–75 will deal in succession with raw materials, manufacture, markets and final distribution, and transportation.

72. The raw materials. The raw materials of commerce fall into three great classes.

1. Those of a mineral nature. Iron is found in certain parts of the world. It exists everywhere in rocks, soils, and even in the waters of the sea, but is so concentrated in certain places as to be available to man. It is in northern and central but not in southern New York; it is about Lake Superior, but not to any extent on the prairies; it is in northern and western but not to any extent in metropolitan England; it is found in Sweden and Spain, but not in Holland. *Where it is*, depends on ancient conditions, and thus geographic distribution leads to geologic origins. Coal has a similar distribution over the world, being widespread but not universal. If salt is taken, or petroleum, or gold, the facts of modern geographic distribution are dependent on ancient conditions, which belong to the geography of some remote time.

. 2. Raw materials of a vegetable nature. Wheat, like all other vegetable products, depends on the existence of soil. This comes from the rocks, and might itself be called a mineral product, belonging to geology as truly as gold or coal. Wheat, however, depends for its actual distribution more on climate,

FIG. 57. Distribution of population in the United States, 1900

Under 6 inhab. per sq. mile
6 - 18 "   "   "
18 - 45 "   "   "
45 - 90 "   "   "
90 and over

that is, on the distribution of temperature and moisture. Hence it does not grow in Iceland or Panama, although those regions have abundant soils. It does not grow in Nevada or Colorado unless water is supplied. It does not grow in the Rockies or other high mountain regions, because altitude introduces semi-arctic conditions.

Cotton has its somewhat distinct zone also, in like manner dependent on climate, and wheat and cotton belong essentially to different belts, although they may overlap, as in Georgia, Tennessee, Texas, or Egypt.

3. Raw materials of animal origin. Cattle will thrive where either wheat or cotton may grow, and thus exemplify the law that a type of animal life will often flourish over more various conditions of soil and climate than are possible to stationary organisms, like plants. Cattle as a commercial product are therefore raised not so much where they might be raised, as where man finds this industry more convenient or profitable than to produce cereals or fruit or cotton. This introduces the human factor.

73. Manufacture. Raw materials are transformed into things which are ready for man's use. Wheat is made into flour or breakfast cereal; cotton is turned into thread, muslin, or lace; cattle are transformed into beef, leather, shoes, harness, and into a multitude of lesser things; iron becomes steel, and steel is given manifold useful forms.

The processes of manufacture belong to chemical, mechanical, or other sciences, which are not in themselves geographical; but they involve many and vital geographical conditions, and thus demand a large place in commercial geography. Thus always the question presents itself, Where is the raw material made into the finished product? The answer to this question is geographical. Still more fundamental is the question, *Why* is the particular place chosen? This involves the locality of origin of the raw material, the possibility of quick and cheap transportation, the presence of power, the neighborhood of supplementary materials, the nearness of the market, the supply

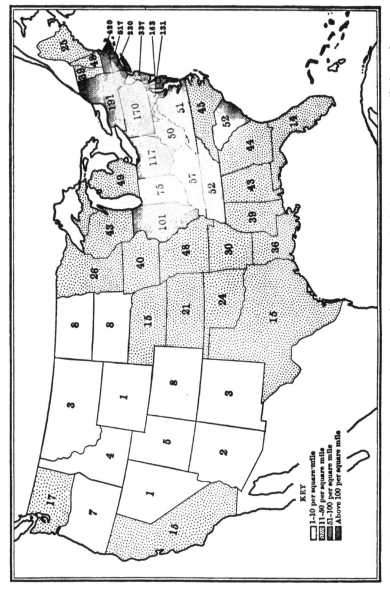

KEY
☐ 1-10 per square mile
▦ 11-50 per square mile
▨ 51-100 per square mile
▓ Above 100 per square mile

FIG. 58. Density of population, 1910. Numerals show number of inhabitants per square mile in each state

98

of labor, the taxes and imposts of government, whether local, national, domestic, or foreign, and also the business skill and inventive genius of the people.

To illustrate some of these points, Pittsburg may be taken as an example already familiar. Power is ready at hand in the form of coal. Iron is transported thither under favorable conditions, but transportation depends upon the genius of man in developing the " Soo," the lake steamer, and the railway. Inventive skill in manufacturing processes has been marked, but is not peculiar to Pittsburg ; for example, the Bessemer process was a British invention. Labor is abundant, because with ready and cheap transportation it comes from the overcrowded countries of Europe. The growth of the Pennsylvania Railroad system and the moderate distance from New York and other Atlantic ports have made it possible to deliver the product where it is wanted. American enterprise, the power of large combinations of capital, and the device of protective tariffs have made it possible to market much of the steel almost at the mouth of European mines. Thus the location of Pittsburg iron making involves a great variety of factors more or less geographical, and belonging both to the physical and the human phases of the science. A similar application of geographic principles covers Birmingham in reference to iron ; Atlanta, Lowell, Fall River, or Manchester (England) in the field of cotton manufacture ; or St. Joseph, Omaha, or Chicago as centers of meat packing.

74. Markets and final consumption. Commercial products find their chief markets among the largest and most progressive populations. Hence all kinds of manufacture seek, so far as possible, the places near which the most people live. A great factory is not often placed in a remote and sparsely populated region. A glance at a railroad map of the United States shows that the closest network of roads is east of the Missouri and north of the Ohio and Potomac rivers. Hence Pittsburg, Buffalo, and South Chicago are more accessible to the market for steel rails than Birmingham or any point in the Cordilleran region. Manchester is a long way from the cotton fields of the South or of

Egypt, but it is close to the vast body of British and other European consumers. There are important exceptions to this rule, for Manchester may market its fabrics in India or Argentina, and its weavers may buy beef from Buenos Aires and butter from New Zealand. Thus centers of population tend to concentrate industry, while transportation favors dispersal, especially where two remote lands can supplement each other, and ships can bear cargoes in both directions.

The location of the market depends not only on the number of people, but on what they want. The use of wheat bread and of cotton is extending in some countries of the world because many people are finding these things desirable. Hence the market and the best locations for manufacture change with the progress of nations and of the race as a whole. Thus China breaks down its walls of conservatism and desires the products of American, British, and German skill. Business enterprise is alert to exhibit modern products and convince primitive or backward nations that they must have them.

75. Transportation. This is a fourth fundamental factor essential to the geography of commerce. Carriage is required at every stage, — from field or mine to the factory, and from factory to the market and the consumer. There may be a series of manufactures, as when steel made in the furnace goes to machine shops of every kind and to factories where agricultural implements, sewing machines, typewriters, or builders' hardware are made.

Thus transportation takes the various raw materials, with pauses en route, all the way to the consumer's door. The place of the intermediate manufacturing and marketing stations depends on a variety of causes. Cotton may be made up near the smaller number of consumers, as at Atlanta ; or far from the field but close to the consumer, as at Fall River or Manchester.

Effectiveness in transportation depends on many factors both natural and human. Take first the natural conditions. River valleys favor those points of production, manufacture, and consumption which they join. The Hudson-Mohawk, the Potomac, the Rhone, and the Rhine are examples. If a mountain range separates two progressive regions, its passes are highways, and are historic if the countries are old. Thus the St. Gothard, the Simplon, and the Brenner passes have long been arteries of trade between central Europe on the north and Italy and the Mediterranean on the south. The Union Pacific Railway finds a low place in the Rocky Mountains of Wyoming, and the Canadian Pacific follows a series of passes from Banff westward in the Canadian Cordilleras, both roads becoming great commercial links between two oceans.

The greater rivers and lakes serve as inland waterways, and to these man adds canals, by which he joins river systems and makes the difficult parts of rivers subject to commerce. He goes farther, cutting the lands in two to join the seas and oceans in unbroken communication around the world. Perhaps even more important than other land forms are plains. Thus railways are cheaply and rapidly built along the coastal plain from New York to Savannah, Jacksonville, New Orleans, and Texas, or along the Great Lakes and across the prairies. The Great Plains are equally favorable, and the grades are imperceptible to the eye, in Kansas, Dakota, Manitoba, or along that more northern line where the Grand Trunk Railway crosses Saskatchewan and Alberta toward Prince Rupert and the western ocean.

More truly than any land, however great, the ocean is the natural highway of nations, and has been tracked by ships as anciently as the trade routes of the Orient have been trodden by hoof of horse or camel.

Seas reach the borders of every continent and island, and their waters are the public possession of nations. The Atlantic Ocean is thus far the great theater of marine commerce, being bordered on the east and west by the greatest commercial peoples. Europe is indented by gulfs and seas, including the diversified outlines of the Mediterranean. North America likewise has the Gulf of Mexico, the Gulf of St. Lawrence, and Hudson Bay. Both continents have secure and plentiful harbors. The

Atlantic has the Gulf Stream and the prevailing westerlies, and toward the equatorial region is marked by the presence of the trade winds and the equatorial current. All these natural features are made useful to the ships of commerce in determining their routes and facilitating their passage. Land and sea routes are more and more coördinated, sometimes under a common management, so that commercial transportation may now be considered as a world-embracing organism.

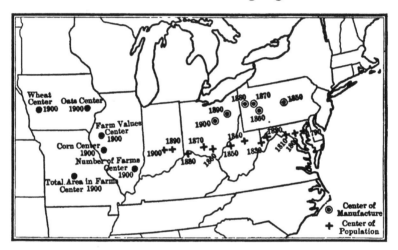

FIG. 59. Westward migration of centers of population and manufacture; centers of oats, farm values, etc., 1900

Transportation is dependent on invention, and the human porter, the pack horse, the freight wagon, the steam and electric car, mark steps of historical advancement on land, as the canoe of the savage inaugurated on the sea a series of craft which has its present culmination in the great liner of the ocean. Invention not only creates the means of mechanical carriage but adds conditions of comfort to the traveler, of security to the cargo, and of preservation for products that are perishable. Much commerce in food materials would be impossible and the future prosperity of some nations might be threatened, but for the device of preservative processes such as hermetic sealing, chilling, and freezing.

76. Complex results of commercial processes. It must not be thought that commerce involves a simple progress from the sources of raw material to final consumption. Every great industry gives origin to others. It carries with it a chain of causes and effects. Cotton is raised and baled with the aid of modern appliances, such as the plow, planter, and machines for ginning and pressing. To make these, other centers and other kinds of industry are fostered. To carry the cotton requires

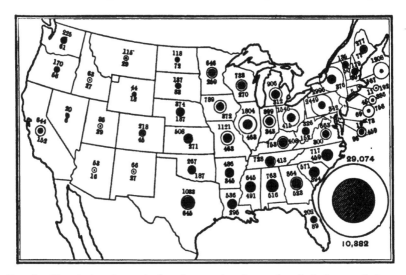

FIG. 60. Population in agricultural pursuits shown by dark inner circle, — number in thousands, below; population in all pursuits shown by outer ring, — number in thousands, above (census of 1900)

railways and steamships, and thus workers in wood, in iron and other metals, decorators and furnishers, are necessary; also expert operators, business managers, and innumerable helpers. To manufacture the cotton introduces the machine maker, capitalist, foreman and operative ; the sales agent, wholesaler and retailer ; in other words, communities are built up and the single industry branches in directions so numerous as to be impossible to trace.

77. Human factors in commercial geography. The foregoing principles are primarily related to the earth and to natural

HAND AND MACHINE LABOR COMPARED

Legend: ▬ = hand, ▬ = machine

Description of unit	Year produced	Number of hours worked (hand)	Number of hours worked (machine)
AGRICULTURE			
Barley: 100 bushels	{ 1829–1830 / 1895–1896 }	211.94	9.04
Corn: 50 bushels shelled, stalks, husks, and blades cut into fodder	{ 1855 / 1894 }	228.35	34.38
Cotton: seed cotton, 1000 pounds	{ 1841 / 1895 }	223.73	78.70
Hay: harvesting and baling 8 tons timothy	{ 1860 / 1894 }	284.00	92.53
Oats: 160 bushels	{ 1830 / 1893 }	265.00	28.39
Potatoes: 500 bushels	{ 1866 / 1895 }	247.54	86.36
Rice: 10,000 pounds rough	{ 1870 / 1895 }	235.16	64.55
Rye: 100 bushels	{ 1847–1848 / 1894–1895 }	251.93	100.67
Strawberries: 500 quarts	{ 1871–1872 / 1894–1895 }	216.54	84.42
Sweet potatoes: 50 bushels	{ 1868 / 1895 }	151.11	58.15
Wheat: 50 bushels	{ 1829–1830 / 1865–1896 }	160.63	7.43
MINING			
Coal: 50 tons bituminous	{ 1895 / 1897 }	171.05	94.30
QUARRYING			
Drilling granite: 60 2¼-inch holes 1⅛ feet deep in granite rock	{ 1897 / 1897 }	178.35	29.64

104

	Year	Value	Year	Value
Granite: quarrying 50 cubic feet	1860	252.00	1896	65.50
Marble: quarrying 72 cubic feet	1876	133.57	1896	26.08
TRANSPORTATION				
Loading grain: transferring 6000 bu. wheat from storage bins or elevators to vessels	1853	222.00	1896	53.60
Loading ore: loading 100 tons iron ore on cars	1891	200.00	1896	2.86
Unloading coal: transferring 200 tons from canal boats to bins 400 feet distant	1859	240.00	1896	20.00
Unloading cotton: transferring 200 bales from vessel to dock	1860	240.00	1896	75.50
MANUFACTURES				
Plow: 1 landside plow, oak beams and handles		118.00	1896	3.75
Shoes: 10 pairs men's fine-grade, calf, welt, lace shoes, single soles, soft box toes	1865	222.50	1895	29.66
Crackers: 1000 lb. graham crackers packed	1858	160.00	1895	35.56
Newspapers: printing and folding 36,000 pages	1895	216.00	1895	1.08
Lithography: printing 1000 sheets art work, 19x28 inches, 6 colors	1867	281.00	1896	5.68
Butter: 500 pounds in tubs	1866	125.00	1897	12.50
Harness: 1 set double coach harness traces, 10 stitches per inch	1860	234.50	1895	40.72

conditions, but even these cannot be discussed without passing over into the human realm. Man is the agent in commerce, and even its geographical features are shaped by him. Theoretically it might be expected that each region would produce what its mines and quarries contain, or what its soil and climate are fitted to yield, and sell its surplus where needed. This would imply equal intelligence, similar standards of living, and equally advanced development of all regions according to their capacity, with unhindered interchange among nations. But the actual facts present a far different picture, and it is the real and not the ideal which here claims attention. The following sections give a partial analysis of some of the human factors.

78. Different standards of living. A market can only be had where the people feel, or can be made to feel, the need of a commodity. Thus American farm machinery makes its way slowly into China or even Russia. The past of wheat is especially interesting because it has gained wider standing as giving the bread of civilization. Its future is the greater problem because it is uncertain how far the demand for the wheaten loaf will go throughout the world. Germany is increasing her consumption of wheat because her working classes do not rely as much as formerly on black breads of rye and barley raised within the empire. The use of tropical fruits is extending among the people of temperate regions. Thus it is but a few years since grapefruit has taken its place beside the orange in the markets of the Northern states.

79. Different degrees of skill. Great resources may be idle, as iron and coal in China, or largely even in Spain. This applies to the winning of raw material, to manufacture, to transportation, and to the organization of commerce in its financial aspects. The North American Indians wrought drift copper, using such fragments as the ice sheet broke from native ledges and distributed with the glacial drift. The white man delves a mile into the earth and brings forth abundant supplies. The savage made a spearhead or a rude ornament. The white man constructs an apparatus for transmitting electric energy for power, heat, or light.

A noteworthy advance is seen if we compare the amount of time required to produce a given result by hand labor and by machinery. The table on pages 104–105 gives such a comparison. It will be seen that in many fields of production a revolution has taken place within a half century, or even within a much shorter period of time.

80. Different stages of occupation of lands. Land is intensively tilled on the plains of the Po or on the flood grounds of the Seine. It is well utilized, but less fully, on the prairies of Illinois, and is barely in the beginnings of culture in western Canada or on the plains of Argentina. In the subduing of the earth there has been a migratory progress, in which men of competent or advanced races find and settle new countries and exploit by their skill and energy new articles of commerce.

81. Division into political units. This has often taken place along lines that are arbitrary and on which artificial restraints are set up, such as tariffs. The sales of a product may be subject to internal taxation, as, for example, tobacco and alcohol in the United States ; or they may be made a government monopoly, as salt and tobacco in Italy. It is said that parts of the Italian coast have been patrolled to prevent the peasants from getting their salt by the evaporation of Mediterranean water. A change of government may introduce new commercial conditions suddenly, as lately in the Philippines and Cuba, or less recently in Alaska, South Africa, and India.

On the other hand, progressive governments greatly facilitate commerce in numerous ways.

Fundamentally, government guards the rights of property as between man and man both on land and sea. Government sets up courts of law and maintains lawful standards of weight, measure, and value. Piracy is suppressed, and at the present time there is growth of governmental activity in preventing unlawful and unjust restraints of trade by monopoly and political interference.

Government affords protection from the hazards of nature to which commerce is exposed. It charts the seas, maintains

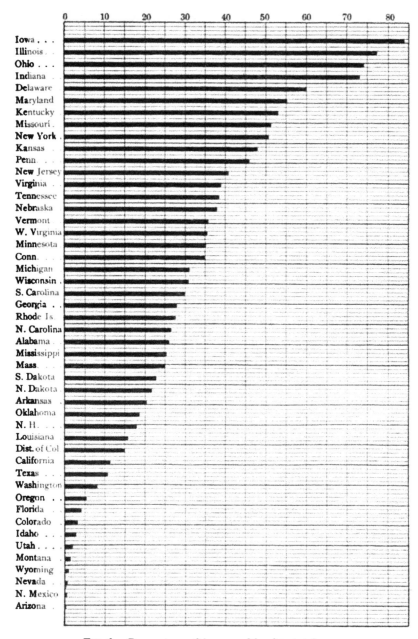

FIG. 61. Percentage of improved land to total area

108

life-saving service, gives weather and flood warnings, maintains lighthouses and. signal stations, and sets up a quarantine. It guards against the migration and importation of pests, as is illustrated by the presence on the plains of Canada of officers whose duty is to enforce laws for the suppression of noxious weeds. The attempted extermination of the gypsy moth by the state of Massachusetts is a further example, as is the stringent control exercised by the Department of Agriculture over the foot and mouth disease by which the cattle and dairy industry is threatened. Under this head falls sanitation, as organized by government authority at Havana, Panama, and Manila.

Government also promotes transportation. It gathers scientific information dealing with routes, practicable roadways, the resources of adjacent territory, and the development of termi- . nal points. The Pacific Railway surveys of the decade 1850–1860 offer an example, and also the land grants made to the Union Pacific and other transcontinental roads in the years of construction following the Civil War. Governments often subsidize lines of marine shipping, and they have given much aid in the construction of canals. River and harbor improvements belong under this class of government activities affecting commerce, as does the extensive development of earth roads and highways. Great Britain, Norway, Switzerland, France, Italy, or indeed any advanced European nation, is conspicuous for its building of roads ; and the United States, though far behind in this particular sphere, owing to the newness of the country, is now extending this important aid to commercial operations. Government mails and telegraphs are further factors in trade.

Governments constantly investigate natural resources, such as mines, soils, and climate. It is enough to name, in the United States, the Department of Agriculture, with its Soil Survey, its Bureau of Plant Industry, its experiment stations, and other activities ; or the United States Geological Survey, its topographic mapping, and its study of general and economic geology. In addition may be named the Reclamation Service, for irrigating the arid lands ; and the Forest Service, for forest protection and the regulation of timber cutting and pasturage. Also here belong the consular services, by which every progressive nation receives fresh information regarding markets, crops, new products, and commercial conditions from all parts of the world for the benefit of the farmer, the miner, the manufacturer, and the trader.

82. Social and economic conditions. Here fall matters belonging strictly in the realm of economics and sociology, which have, however, important bearings on the activities of commerce. Such are the restraints imposed by trade unions and by monopolistic combinations of capital. Here may be named again government subsidies and bounties, which interfere with the more appropriate productions of a region or with the natural routes of commerce. Cities having every natural advantage may be retarded by the retrogressive spirit of the people or the repressive policies of labor or capital; or they may be brought to prosperity by special favor, in the face of natural limitations. It may well be doubted, however, whether a great and permanent center of commerce can arise, if not favored both by nature and by man.

PART II. COMMERCIAL GEOGRAPHY
OF THE UNITED STATES

CHAPTER VII

PHYSICAL FEATURES OF THE UNITED STATES

The student who begins commercial geography should have mastered the chief principles of physical geography and should know his own country well. This short account of the United States will not supply such knowledge. It will serve to review a few leading facts and to give a bird's-eye view of American geography in its relations to commerce.

83. Position. The lands of the republic are in the middle latitudes of North America, and, excepting Alaska, include no frozen grounds or Arctic wastes; nor are there tropical lands, save the island possessions gained in recent years. The continental United States has enough land in regions of temperate climate to provide all characteristic grains, vegetables, and fruits; and it reaches so far toward the tropic of Cancer as to produce abundance of cotton, rice, cane, and subtropical fruits. Being on the Gulf of Mexico and near the Caribbean Sea, this country readily trades with the West Indies, Central America, and the tropical parts of South America.

The nations which led the civilization of modern times are in western Europe. The lands of the United States are directly across the Atlantic and were naturally colonized by Spain, France, Holland, and Great Britain. In more recent times almost every country of Europe has given many of its people to America, so that ancestry, present kinship, and neighborly positions on

FIG. 62. Relief and drainage of the eastern United States

opposite shores of a narrow ocean have created on the North Atlantic Ocean the chief trade of the world. Having a long frontage on the Pacific, this country readily reaches out to Japan, China, and the newer countries and islands of that ocean. A commercial position of this kind is held by no other country save Canada.

84. Atlantic region. On the Atlantic border is a broken shore line whose bays and tidal rivers once tempted the explorer and pioneer, as they now invite the ships of commerce and form the gateways to great cities. The shallow waters of these inlets extend from 50 to 100 miles out from shore over the continental shelf, and here the fisherman plies his trade, and the oyster-man gathers his harvests from the Chesapeake to the Banks of Newfoundland.

The Atlantic lowland from New York to Eastport, Maine, is a worn mountain land, once lofty and rugged, now reduced and merely hilly, bearing the farms, towns, and cities of coastal New England. From New York to Florida the lowland is a coastal plain cut by many tidal rivers and by deep bays such as the Delaware and the Chesapeake, and stretching westward to the fall line and to the cities of Philadelphia, Baltimore, Washington, Raleigh, and Columbia. These cities mark the passage from the smooth coastal plain and its sluggish rivers to the ancient foothills of the Appalachians. The Atlantic region shows a fiord coast. It is without the rugged heights of the fiords of Norway or Scotland, and in the south is more like eastern England, with the " drowned " lower courses of the Thames, the Humber, and the Tyne. Here, therefore, are safe harbors, good sites for cities, valleys offering roadways to the interior, and fertile fields lying close at hand. All these conditions are helpers to commerce.

85. Appalachian highlands. Rising as a background from these eastern lowlands are the Appalachian highlands. These are mountain ranges and plateaus of various ages, shapes, and heights. As a whole they stretch from northeast to southwest, with Maine at one end and Alabama at the other. To them belong the highlands of Maine, the White Mountains, the Green Mountains, and the Berkshires of New England, the Adirondacks and the Highlands of the Hudson. Southward they include the Blue Ridge and the long parallel ranges that reach from eastern Pennsylvania to central Alabama. West of these mountains is the Appalachian plateau, more familiarly known as the Catskills in New York, the Allegheny uplands in Pennsylvania, and the Cumberland plateau in Kentucky and Tennessee, and reaching out toward the Mississippi River in Ohio and Indiana.

These heights are crossed by important valleys, of which the following have the greater commercial value : the Hudson-Champlain in New York and Vermont; the Hudson-Mohawk in New York ; the Susquehanna in Pennsylvania ; and the Potomac in Maryland and Virginia. To meet the more southern of these openings the Ohio and its branches reach into the highlands from the west. By way of these gaps the earliest railways and canals in America were laid out, and here to-day are found the New York Central, Pennsylvania, and Baltimore and Ohio railway systems. These passes likewise have determined the position of New York, Philadelphia, and Baltimore, and will, through all time, contribute to their growth and sustain their trade. ·

Throughout most of the length of the Appalachians runs a broad valley, fertile in soil, rich in towns, and serving as a great highway. Parts of this great Appalachian valley are known as the middle Hudson valley in New York, the Cumberland valley in Pennsylvania, the valley of Virginia, and the valley of East Tennessee. Among its towns are Reading, Harrisburg, Hagerstown, Staunton, Knoxville, Chattanooga, and Birmingham.

The Appalachians have forests of pine, spruce, and hard woods, offer fertile valleys running among the heights, are abundant in water power, are rich in iron and coal, and hold these resources within easy grasp of the great commercial centers and smaller industrial cities of the Atlantic lowlands.

86. Mississippi Basin. Unlike as they look on the map, North America and Europe resemble each other in structure and general relief. Europe has its lower and older mountains in the north, its younger and higher mountains in the south. We have lower and older heights in the east, younger, loftier, and more rugged mountains in the west. Lying between the mountains in each continent is a lowland. Holland, North Germany, and Russia belong to the central lowland of Europe, as the Mississippi basin and the basins of the Nelson and Mackenzie rivers form the vast interior lowland of North America.

Comparing the Mississippi Valley and the great plains of Russia, there is much likeness

Prairie
Woodland

FIG. 63. Original distribution of prairie and woodland in Illinois (Barrows)

in surface, size, and fertility, but the American plain is better placed commercially. Russia has outlets on the Black and the

Baltic seas and the Arctic Ocean, and its greatest river flows into a closed sea, while the Baltic ports are often shut by ice. The Mississippi Valley has its outlets by the great railways to both seaboards, by the Great Lakes to the Atlantic, and by the river to the Gulf of Mexico, and soon, by the Panama Canal, will have access to the Pacific Ocean.

In extent of rich and productive soil only two other regions in the temperate zones can be compared with the Mississippi Valley. One is the European lowland already named, and the other is eastern and central China.

The Mississippi basin includes the western slope of the Appalachian highlands, as already described. On the west, in the states stretching from the Dakotas to Texas, lie the so-called Great Plains, really a plateau sloping eastward from the Rocky Mountains. Thus two plateaus descend toward the Mississippi River, but before they reach it they pass into the prairie lowlands of Iowa, Illinois, and other central states. The Mississippi and the prairies are in a region of sufficient rainfall. The rainfall decreases westward until deserts lie at the foot of the Rocky Mountains. It increases eastward until heavy forests of hard wood crown the heights of the Appalachians. With suitable improvement of the rivers nearly all parts of this region can have transportation by water, and its surface has ever favored the construction of railroads.

87. Gulf region. Southward are the Gulf lowlands, consisting in part of the delta of the Mississippi, and in greater part of coastal plain, some of which is prairie, some forest, and all a land of productive soil, favorable climate, and open by land and water to the operations of commerce. The Panama Canal will place this southern belt near one of the trunk lines of the world's commerce.

For the purposes of trade no line can be drawn between the Mississippi Valley and the regions bordering the Great Lakes. The divides between the two systems of waters are so low as to make the passage between them almost imperceptible. Thus Minneapolis connects with Duluth, the valley of the Wisconsin

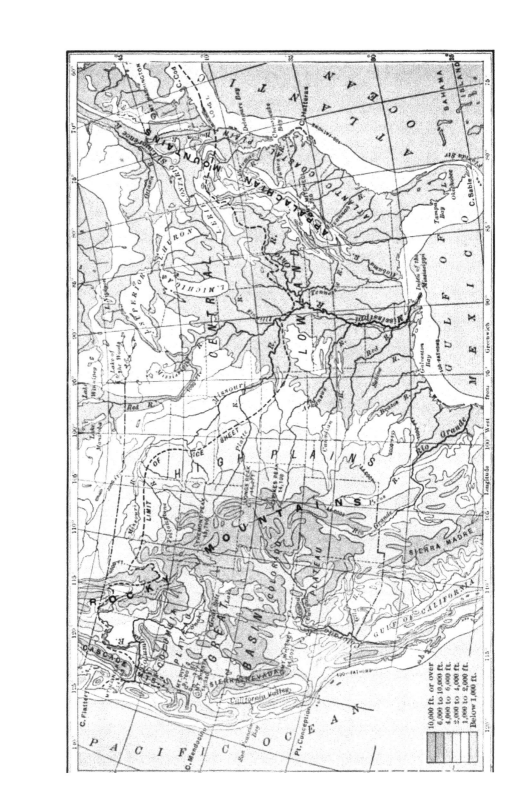

River with Milwaukee, and Illinois and surrounding states with Chicago. Around the lakes are extensive arable lands in Michigan, Wisconsin, Indiana, Ohio, and New York. The lakes themselves afford a highway to the eastern seaports not only of the lower St. Lawrence, but, by the Erie Canal and the various railways, to New York. The lakes were once larger than now, and some of the lands are plains representing the ancient lake bottoms. Such also are the rich prairies of the Red River valley, both in the United States and north of the border, in Canada. Some of the lands about Lake Superior are ancient mountain country, now greatly worn down and holding stores of copper and iron.

88. Cordilleran highlands. The routes of settlement and many of the highways of commerce cross the Atlantic lowland, pass the divides of the Appalachians, and descend to the Great Lakes and the Mississippi River. Thence they rise across the Great Plains to the base of the Rocky Mountains. Here they enter the western uplands, which are properly called the Cordilleran region. The first member of these uplands is the Rocky Mountain ranges, extending from Montana and Idaho through Wyoming, Colorado, and Utah. Rising to heights of two or nearly three miles above the sea, they catch abundant moisture and send down their streams to water the arid plateaus both east and west. They are most massive in Colorado, but the railways to the Pacific find passes across them in every state through which they extend.

West of the southern Rocky Mountains are the Colorado plateaus, grand in physical features, especially in giant canyons, but poor in natural resources. Parts of Wyoming, Utah, Colorado, New Mexico, and Arizona are included in this region. West of the northern Rocky Mountains are the lava plateaus of the Snake and Columbia rivers, an arid region, but capable of production where water is available for irrigation.

Beyond the Colorado plateaus, in the western half of Utah and in Nevada, is the Great Basin, which is one of the driest parts of the western country. It is bordered by the lofty Wasatch

Mountains on the east and the bold Sierras on the west. The best supply of water is from the Wasatch, and here, since the beginning of the Mormon settlements, the hand of man has made a broad oasis in the desert.

89. Pacific coast region. The coast states will be best understood by following the course of the Sierras, which pass through eastern California and continue northward as the Cascade Range through central Oregon and Washington. Above the forests of the Cascades rise snow-crowned volcanic cones, as sentinels looking out upon the Pacific Ocean. Following the shore line through the three states are lower and younger mountains. These are still, it would seem, in the process of building, as was shown by the extended dislocation and attendant earthquake shocks of 1906. These mountains bear the names Coast Range in California, Klamath Mountains in Oregon, and the Olympic Range in Washington.

Between the greater range on the east and the lesser on the west is a series of broad and rich valleys. These are the central valley of California, with its fields of grain and fruit; the fertile lowlands of the Willamette valley in Oregon ; and the plains extending along Puget Sound and southward in Washington.

With much of rugged mountain and intervening desert, the Cordilleran region has great resources. The area under the plow will never be large, but the culture will be intensive, as scientific irrigation makes good use of the water supply. And as the mountains of the East have wealth in iron and coal, so the western mountains and plateaus supply gold, silver, lead, mercury, and copper, and the climate favors those human energies which make the most of every resource.

The harbors of this shore line are not many as compared with the Atlantic, but each state has one goal for Pacific shipping. San Francisco stands at the Golden Gate, where the waters of the valley of California pass into the ocean. The lower Columbia River leads up to Portland in Oregon, and Seattle and Tacoma stand on the shores of Puget Sound. Lacking in the other coast states, coal is abundant in Washington, and, in compensation for the deficiency, California has become a large producer of petroleum.

No parts of the United States show a greater variety of climate or surface than the three great states of the Pacific coast. They combine regions of large rainfall and districts so arid that crops can only be raised through the aid of irrigation. Broad and fertile valleys invite the raising of all kinds of fruits and grains, while the mountain slopes and plateaus bear the greatest of American forests. In response to these natural advantages this region has in recent years seen large growth in its cities and great expansion of its industry and commerce.

Thus in the United States, as in all lands, mountains and plains, lakes and streams, the soils and the rocks beneath, the moisture of the atmosphere and the form of the shore lines, help to shape production, to locate the centers of trade, and to trace the lines along which man and his merchandise move.

CHAPTER VIII

PLANT PRODUCTS OF THE UNITED STATES

90. Wide range. Plants depend on soil, moisture, and temperature. The United States, from Florida to the Canadian boundary, covers 24 degrees of latitude, and hence its climate varies from subtropical to temperate. Therefore cotton, sugar cane, and tropical fruits flourish in the southern parts, and wheat, oats, barley, apples, and Irish potatoes abound in the north. Corn is intermediate in its character, and may be grown from Wisconsin and New York to Louisiana and Georgia. The country also embraces 58 degrees of longitude, and has great variety of surface from the Atlantic and Gulf plains to the Appalachian uplands and the heights of the Rocky Mountains and the Sierra Nevada. With these features is coupled a large range in rainfall, from the deserts of the Great Basin to the plentiful moisture east of the hundredth meridian, and the heavy precipitation of the southern Appalachians and of the northwest coast.

The consequence of these conditions is that nearly all the vegetable products needed by man can be raised on our own soil. Canada, in contrast, ranges from temperate to arctic, and must import many necessary things. No European country has so large a range, though Italy, Spain, and France, by virtue of variety of surface, go far to make up for their lesser extent of latitude. It will be seen that the range of vegetable products in the United States steadily widens by the importation and breeding of new plants and by irrigation and more scientific farming. We tend therefore, so far as natural resources go, to become a self-sufficient nation.

Plant products fall into a number of groups. Food plants include cereals, fruits, vegetables, sugar-producing plants, plants yielding beverages, and indirectly those used for forage. Fiber plants and forest products show a vast variety of uses, though the more important of these is protective, in the way of clothing and shelter. These groups will be taken up in their order.

91. General view of cereals. Cereal plants produce edible grains and strictly belong to the family of grasses. For convenience other grains not grasses may be included. Buckwheat is an example. The importance of cereals is shown by the fact that in 1899 (Twelfth Census) the ground occupied by them was nearly 64 per cent of the area occupied by all crops. The states of the North Central division led all others. This division includes the chief prairie states and extends from Ohio along the Great Lakes and includes on the west the Dakotas, Nebraska, and Kansas.

Oats showed the greatest yield per acre (31.9 bushels), followed by corn (28.1) and barley (26.8). Wheat and rye were at the bottom of the scale, with yields of about 12½ bushels. In

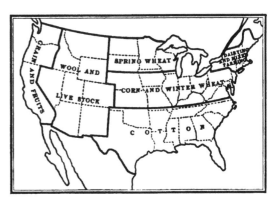

FIG. 64. General distribution of the more important agricultural products

order, the leading cereal states stood as follows at the time of the Twelfth Census, whose figures are for the year 1899: Iowa, Illinois, Kansas, Nebraska, Minnesota, Missouri, Indiana, Ohio. The total production in the United States was nearly 4,500,000,000 bushels. The production for 1908 was 4,376,-000,000 bushels, this figure including seven grains in order, from highest to lowest: corn, oats, wheat, barley, rye, rice, and buckwheat.

The geographic center of production of six cereals, omitting rice from the above list, was in 1900 on the Mississippi River between Quincy, Illinois, and Keokuk, Iowa. In 1850 the center was near Cincinnati, and has thus moved westward, and in less degree northward. From 1890 to 1900 the center moved westward but 25 miles, due to the fact that the new lands of the West had been mainly taken up. At the same time (1900) the center of population for the United States was in Indiana and the center of manufacture in Ohio. Thus there is a greater bulk of manufacture east, as there is larger raising of cereals west, of the center of population.

92. Cereals of the temperate regions. Wheat has been studied in Chapter I. It is the most important of cereals, although corn and oats surpass it in amounts produced.

Corn has a special importance for the United States because it is a native of the western world and because the climate is favorable to it. It was readily raised by the pioneer on fields from which the trees had been cut, and was ground for his food on the top of a stump or in primitive gristmills. While wheat is a native of the Orient, it has been brought to the new continents. Corn in like manner has become a grain of many nations, but retains its center here, both in production and consumption, being used as food for man but much more largely for raising stock. Americans restrict the term "corn" to maize, while the English apply it to wheat and cereals in general.

Corn well illustrates the importance of climate in commercial geography. Some corn is raised throughout the latitude range of the United States and in southern Canada, but the large production extends from central Michigan and Wisconsin into the Gulf States. Within this region, however, is a narrower area known as the "corn belt," with Illinois and Iowa as the most important states, and including Ohio on the east and Nebraska and Kansas on the west. The belt lies mainly between 39 and 43 degrees north latitude.

Corn cannot endure frost; hence the season must be long enough to protect both the young and the maturing stages. It requires hot weather during its period of growth, with ample supplies both of rain and sunlight. These conditions are well provided in the prairie region of the Central States. In like manner corn flourishes in Roumania, Russia, and the Mediterranean countries, but cannot be grown in Great Britain or the northern parts of the European continent.

The corn crop of the United States in 1908 was 2,558,000,-000 bushels, of which Illinois produced 298,000,000 bushels, Iowa 287,000,000, and Nebraska and Missouri each a little more than 200,000,000 bushels.

Oats afford the second largest cereal crop in the United States, but in commercial importance are much inferior to wheat, the

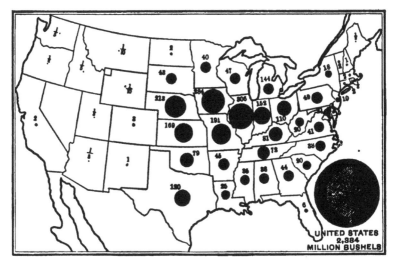

FIG. 65. Average annual production of corn in millions of bushels, 1899–1908

trade in them being chiefly local and domestic. In 1850 New York, Pennsylvania, and Ohio produced the most oats, but in 1899 four fifths of the crop were raised in the North Central States, while Illinois, Iowa, Wisconsin, and Minnesota produced more than one half of the total.

Comparing the chief cereals as regards their general range in north latitudes, oats, rye, and buckwheat are more distinctly suited to northern localities, while rice and Kafir corn flourish in southern or at least warm temperate regions. Wheat and corn are somewhat intermediate, wheat inclining to the north and corn to the south. Barley has a wider climatic range than any other cereal. It is grown in Norway at 70 degrees north latitude, and in the northern parts of Russia and Siberia, and it flourishes also in the subtropical regions about the Mediterranean.

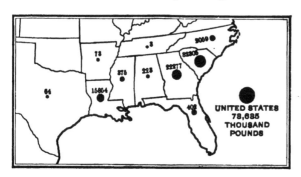

It is not a bread cereal in the United States, being chiefly used for malt liquors. The production in 1908 was only 166,000,000 bushels. The chief area in the United States is between Lake

FIG. 66. Rice production in thousands of pounds, 1869 (Census of 1870)

Michigan and the Missouri River, with minor areas in western New York, about the head of Lake Erie, and in California.

Rye is a much less important crop than barley, amounting to 31,000,000 bushels in 1908. It is used chiefly for distilling whisky. Like barley, it is an important food crop in northern Europe. It is grown in New York, Pennsylvania, southern Michigan, and Wisconsin. Buckwheat was produced in 1908 to

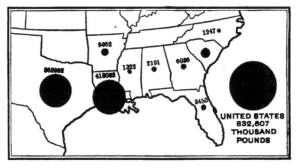

FIG. 67. Average annual production of rice in thousands of pounds, 1904–1908

the extent of 15,000,000 bushels, and, unlike any other grain in the group studied, keeps its center in the North Atlantic region.

93. Cereals of subtropical regions. Rice is the most important cereal in the world, if the numbers which it feeds be taken as the criterion. Judged by standards of living, and having regard to the most advanced peoples, wheat is the chief cereal. Rice is a minor product in the United States, amounting in 1908 to 21,000,000 bushels; but its culture is increasing. Being a native of the East Indies, it finds similar conditions in the South Atlantic and Gulf regions. It is grown in every coast state from Virginia to Texas, and on small areas in Arkansas. Louisiana ranks first in the production of rice, South Carolina second, and these are followed by North Carolina and Texas.

FIG. 68. Harvesting rice, Louisiana

Suitable land is found in marshes and river bottoms, tide flats, the prairies of Louisiana and Texas, and on some uplands. For a long period rice was the only crop in the United States cultivated by the aid of irrigation. The most important localities for the past 10 years have been southwestern Louisiana and eastern Texas, where modern methods of irrigation have been adopted and immigrants have come from parts of the Mississippi Valley to the northward. The change from the flail of the old days and from winnowing in the wind, to steam threshers, cleaning from 1200 to 1500 bushels in a day, illustrates the progress of the industry.

Kafir corn is a native of Africa, as its name suggests. It is related to sorghum, and was introduced in recent years by. the Department of Agriculture as a substitute for Indian corn. It is raised in Kansas, Oklahoma, and other parts of the Southwest, being adapted to a hot and dry climate. Its ability to meet droughts makes it productive where Indian corn is likely to fail. In 1893 the area of its production in the United States was less than 50,000 acres. In 1899 more than 600,000 acres were devoted to this grain.

94. Subtropical fruits. The great state for these fruits is California, whose product at the time of the Twelfth Census

FIG. 69. Date palms in bearing, six years after planting, Tempe, Arizona

exceeded $7,000,000 in value, no other state reaching even $1,000,000. Florida reported a value of $945,000. It was here that the Spaniards, about the year 1562, planted orange trees in America. Arizona and Texas were the only other states which produced fruits of warm climates to the amount of $10,000. California produces nearly all the lemons raised in the United States. Pineapples, on the other hand, are scarcely grown in this country outside of Florida. Five million pounds of olives were produced in 1899, — all in California. The area of olive orchards in 1908 was 30,000 acres. Figs are produced in California and Texas, and in a small way in all the Gulf States. Date palms have been introduced by government experiment stations in Arizona and California, and it is hoped to lay the foundations of an important industry.

95. Fruits of temperate latitudes. These include by far the greater part of the total fruit product. California produced values of $28,000,000 in fruit in 1899, three fourths of this amount being credited· to fruits of temperate climates. California is the first fruit state, and is followed by New York, Pennsylvania, Ohio, and Michigan. The fruit centers are therefore in the Northeast and Southwest, embracing a wide range of climate.

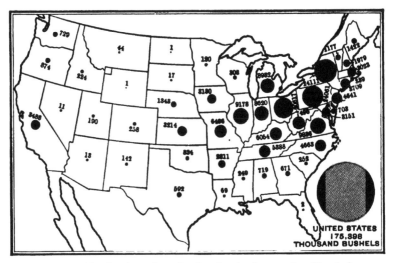

FIG. 70. Production of apples in 1899 in thousands of bushels

The apple is first among American fruits. Trees to the number of 120,000,000 were reported in 1890, and in 1900 the record rose to 200,000,000. New York, Pennsylvania, and Ohio are the greatest producers. Taking all orchard fruits into account, California was first and New York second. These states hold the same order in grape culture, while in small fruits New York leads the list and California is in the eleventh place. California in the last census reported three times as many grapevines as any other state, while New York was second and Ohio third. On the borders of Lake Erie, in Chautauqua County, New York, is a region known as the Chautauqua grape belt. The temperature is moderated by the presence of water, and the soils are in part the silts of glacial lakes which stood at higher levels than the present waters. This single county has nearly 12,000,000 grape-vines and a fruit product of $1,620,000, nearly all of which is from vineyards. Niagara County, similarly situated on Lake Ontario, reports a crop valued at $1,184,000, chiefly apples, peaches, pears, and plums.

The peach is next to the apple in orchard fruits, and is able to take a wider range of climate, being extensively grown in New York and Michigan, as well as in Georgia, Alabama, and Texas. New Jersey and Delaware, once known for this fruit, have fallen off in production, and in a recent year California raised more than half of the product of the United States.

Of small fruits the strawberry is the most important. The cranberry, though not a large product, shows an interesting con-centration in the marsh lands of Wisconsin, New Jersey, and Cape Cod, Massachusetts, — three localities which yield more than 96 per cent of the entire crop.

96. Vegetables. According to the Twelfth Census, vegetables occupied 2 per cent of the acreage of all crops, but made up 8.3 per cent of their value. This means that the land is both more intensively cultivated and that the nature of the crop makes larger values possible on small areas. The states from New York and Pennsylvania to Illinois and Wisconsin were the chief pro-ducers, as might be expected when it is remembered that these states contain several of the great cities, and that the trade in vegetables, owing to their bulk and their perishable character, is in a special degree local.

The Irish potato, so called, is believed to be a native of Chile, and is the chief among American vegetables. It is especially a Northern crop, and Maine, New York, Michigan, and other states on the Canadian border have recently increased their pro-duction, while several Central States just southward have fallen off. Thus there has been a northward movement of the plant. There is a smaller consumption in the Southern states, doubtless owing to the presence of the sweet potato in that region. New

York has been first in this crop since 1850. The annual consumption in the Northern states is about four and one-half bushels per person.

Potatoes are difficult to transport; hence the sale is local and large crops glut the market and cause low prices. A small crop is therefore usually more profitable to the farmer than a large one. The sweet potato ranks second among the vegetables, and is most largely grown along the coastal belt from Virginia to

FIG. 71. Digging potatoes near Greeley, Colorado

Alabama. In small plantings, however, this vegetable covers a surprising range, for the only states that did not report sweet potatoes in 1900 were Maine, Montana, and Wyoming. The effect of climate is well shown in the following comparison. The combined states, Illinois, Indiana, and Ohio, reported sweet potatoes from 64,067 farms, while Michigan, Wisconsin, and Minnesota reported them from 309 farms. A great variety of minor vegetables is grown in truck gardens for market and in family gardens for private use. Their chief interest in commercial geography is that they supply an example of the concentration of industry about centers of population.

97. Preservation of fruits and vegetables. The drying of apples and berries was practiced by the American Indians and by the early settlers. It is a process of commercial importance, — raisins, prunes, and currants being examples. Canning, with hermetic sealing, is now the great method of preserving these otherwise perishable products. Effective beginnings were scarcely made until the period of the Civil War (1861–1865). During the years following, canned goods were a luxury, but are now in

FIG. 72. Potato cellar near Greeley, Colorado

common use among the rich and the poor. The value of canned fruits and vegetables in the year 1905 in the United States was more than $70,000,000. Corn, tomatoes, and peas are the chief vegetables canned, but many others of growing importance appear in the list. The canning process enables these products to pass beyond the local market and enter· in a large way into commerce.

98. Transportation of fresh fruits and vegetables. This trade, which seems familiar and established, is nevertheless young. It is but little more than a half century since the first garden truck was carried by water from Virginia to New York City. Shipments by rail began from Virginia as late as 1885, and Florida

began to send oranges by rail to New York in 1888 and strawberries in 1889. Strawberries were sent to Northern markets in
1875 from Mississippi and Tennessee. In 1888 ripe cherries
and apricots were successfully sent across the continent from
California in refrigerator cars. Of such cars it is believed that
no less than 61,000 were in operation in this country in 1901.
This number includes cars engaged in carrying meat, beverages,
and dairy products, as well as those of the garden and orchard.
Solid trains thus laden now pass from California to the Atlantic
coast, and the car is supplemented at many points by the cold-
storage warehouse.
These arrangements
not only expand the
market in a large
way, but they pro-
long the period of
supply of a given
product. Thus a
resident of New
York has the bene-
fit of successive

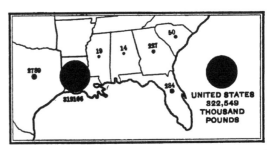

FIG. 73. Production of cane sugar in thousands of
pounds, 1899 (Census of 1900)

ripenings of strawberries while the season of ripening passes
from Florida up the coast to New England.

99. Sugar-producing plants. These are the sugar cane, the
sugar beet, and the hard-maple tree, sorghum being grown for
sirup. Sugar cane has been raised in Louisiana for a century
and a half, and that region is still the chief locality for this
product in the United States. As sugar cane is a tropical and
subtropical plant, the Gulf States on the northern rim of the
equatorial belt are suited to its culture. The domestic product
has never been large, and as the people of the United States
consume much sugar, they import heavily from tropical regions
and now draw supplies from the territory of Hawaii.

Sugar from beets marks one of the younger among the great
industries of the world, but has now come to vast importance,
particularly in Germany. Since 1889 more than one half of the

world's supply of sugar has been derived from beets. In 1853–
1854 beets afforded but one seventh of the total. The first successful factory in the United States dates only from 1870. Ten
years later the product was valued at $250,000. In 1905 the
value had risen to $24,000,000, and there were 51 establish-
ments, of which Michigan had 19, Colorado 9, and California 5 ;
Colorado, however, leading in value of product. The by-products
are not unimportant, since the pulp is used to feed stock, the

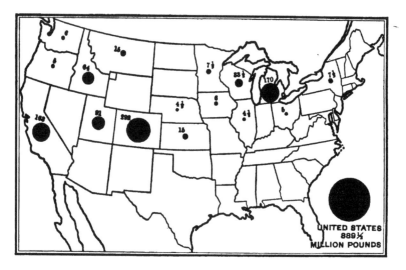

FIG. 74. Average annual production of beet sugar, 1907–1908, in
millions of pounds

juice may be employed for making alcohol, and the lime cake
from some Michigan factories is mixed with asphalt and made
into paving blocks.

In Colorado and Utah sugar beets are grown by aid of irrigation, and the latter state supplies an interesting example of local
transportation. Slicing stations are built 12 to 25 miles away
from the factory, the juice is extracted and forced to it through
pipe lines. Taking the sugar of the world as a whole, we see the
phenomenon of an industry largely migrating from tropical to
temperate regions. It is therefore reasonable to expect that the

FIG. 75. Sugar factory; sugar beets at Blissfield, Michigan

United States will here take a step in commercial independence by producing in the near future its own sugar.

100. Plants yielding beverages. Reference has been made to some of these in the account of cereals. Corn and rye serve for the distillation of whisky, and barley malt for the brewing of beer. Wine making has no such relative importance as in the Latin countries of Europe, but is growing with the expansion of vineyards in California, New York, and other states. According to

FIG. 76. Tea bush, Ceylon type, Summerville, South Carolina

the Twelfth Census the total value of vinous liquors was $6,500,000, while distilled liquors amounted to $96,700,000, and malt liquors reached the great total of $237,000,000. In contrast France has 1,600,000 wine growers, and the vintage for 1907 was valued at nearly $200,000,-000. Apples are used for the production of cider, and hops for the making of beer. Hop raising tends to localization. For 50 years previous to 1890 a few counties of central New York raised most of the hop crop of the United States. In later years the greater development has been transferred to Oregon, California, and Washington. Wisconsin grows a few hops, and the production of all other states is trivial.

FIG. 77. Tea factory at Summerville, South Carolina

For the nonalcoholic beverages, tea and coffee, the United States relies on importation. There is reason to expect that tea may be grown at home, and the Department of Agriculture has already made successful experiments in Texas and South Carolina.

101. Tobacco. This plant has a wide climatic range through the torrid and temperate zones. It has been raised in Scotland, and flourishes in Sumatra, India, Brazil, Cuba, and Mexico. In the United States it grows from Wisconsin to Louisiana and is widely distributed in the East. Richmond is the largest center of tobacco trade in America.

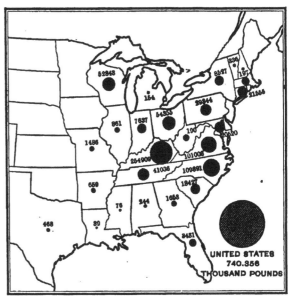

FIG. 78. Average annual production of tobacco in thousands of pounds, 1900–1908

Tobacco is a native of America, but became an important export as early as 1676, and was used as legal tender in Maryland in 1732, the year of the birth of Washington. In his first administration (1789–1793) tobacco was exceeded by flour alone as an export product, and its cultivation and use have gained a surprisingly wide distribution throughout the world. The study of its history is a most useful exercise in commercial geography.

The United States produces far more tobacco than any other country. The crop of 1908 was valued at $74,000,000, and the manufactured output, including some imported raw material, at $331,000,000. In the same year Kentucky was as usual the leading state in acreage, though North Carolina slightly

surpassed her in the amount grown. The third state was Virginia.
A tobacco map of the United States shows that the production
is east of the hundredth meridian, except in a small area in
California. All territory averaging more than 1000 pounds per
square mile is east of the Mississippi River, and the densest
areas lie along the Ohio River, in the Carolinas, Virginia, and
Maryland, and across Pennsylvania to Lake Ontario in New
York. There is also an important area in the Connecticut valley.

FIG. 79. Harvesting alfalfa, Port Conway, Virginia, on the farm which
was the birthplace of President James Madison

102. Forage products. Of these, hay is much the most important, timothy grass and clover being the main sources of this
product, whose value in 1908 was $635,000,000. Forage crops
occupy more than one fifth of the land used for crops in the
United States. At the time of the Twelfth Census they occupied one third as much land as the cereals, and the values were
about in the same ratio, while in 1908 the area was about one
fourth that of cereals. In the North Atlantic division this crop
is more important than all cereals combined, but the North
Central States are first in total value of the product.

The trade is chiefly local, and the crop is related to the dairy
industry of the East and to the cattle raising of the West, with

small exports of baled hay. Hay includes wild, salt, and prairie grasses, millet, and alfalfa, which is related to clover and is especially important in the western parts of the country. Corn, Kafir corn, and sorghum, cut in a green state, are the principal forage plants, much corn being cut before ripening for ensilage.

103. Fiber plants. These include cotton (studied in Chapter II), flax, hemp, wood, straw, and cornstalks. This use of wood

FIG. 80. Flax at harvest time, Pigeon, Minnesota, 1909

will be considered under the head of Forest Products. While flax is here named as a fiber plant, its chief commercial product in the United States is the seed, used for linseed oil, and this industry the United States shares with Russia, British India, and Argentina. The linen and linen fiber used are chiefly imported, for their better quality and their cheapness. The competition of cotton has been too strong for native flax. The seed is raised chiefly in Minnesota and the states bordering the Missouri River. Hemp is a minor crop in the United States, Kentucky being the principal producer. Manila hemp and jute are imported in large quantities, for purposes for which the native fiber might be used. Cotton takes its place for twines and yarns, and the industry has dwindled more than 90 per cent since 1859.

104. Forests of the United States. When the white man came to America he found a dense forest cover stretching from the Atlantic coast somewhat beyond the Mississippi River. There was one important exception in the prairies of the upper Mississippi, but even these were not wholly treeless, for there were many patches of forest, especially along the watercourses (fig. 63).

FIG. 81. Hemp at harvest time, Lexington, Kentucky, 1907

Heavy forests extended from Maine westward across New York and along the Great Lakes far into Minnesota. To the southwest the forest was continuous along the Appalachian highlands to the Gulf of Mexico. Trees grew abundantly along the Ohio and Tennessee rivers and on the Gulf plains, as well as across the Mississippi in the higher lands of Arkansas and Missouri.

These eastern tree growths are often called the hardwood forests, because they are made up largely of such deciduous species as the maples, oaks, chestnut, ash, beech, birch, hickory, and black walnut. Many coniferous trees, however, are found, especially along the Appalachian Mountains and in the Northern states from east to west. Throughout this entire range the white pine is or has been of the first importance, but the supply has in recent years been depleted. The hard or Georgia pine is a Southern species ranging from Virginia to Texas. The cypress is another characteristic Southern conifer. Among Northern conifers, in addition to pine, there are spruces, firs, cedars, and larches, and in some localities the hemlock is abundant.

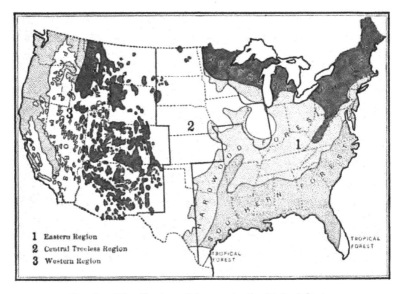

1 Eastern Region
2 Central Treeless Region
3 Western Region

FIG. 82. Distribution of forests in the United States

The Cordilleran woodlands are less extensive and more broken than those of the East, and are chiefly coniferous in character, forests being often made up largely of one or two species of trees of great size. Most important of these is the Douglas fir, which is found extensively in the Pacific coast states and British Columbia, and to a less degree in the Rocky Mountains. Several species of pine are found in the Cordilleran highlands, and in California two species of Sequoia, the redwood and the big tree. Large areas of the Great Plains and of the intermontane plateaus are so arid as to be destitute of trees.

105. Lumber and lumber centers. Lumber, in boards and other forms, is the most important product of the forests. Thousands of sawmills are located at points to which the logs can conveniently be brought. In the Northern states the first movement is over the winter snows. The logs are piled beside streams, down which they are " driven " in the flood water of the spring. Often they are bound into immense rafts and towed across lakes or down the greater streams. The mills are run by water and by steam, and the waste wood in the latter case furnishes a convenient fuel. Sometimes, as at Minneapolis, a great sawing center develops, and a chief market for lumber as well. This special center has water power, is near the forests, and is on the edge of a great prairie, which offers a market. The same is true of other towns farther down the Mississippi. The Great Lakes also offer a highway for the lumber of their bordering states, and hence Ashland, Milwaukee, Chicago, Bay City, Detroit, and Tonawanda, New York, are lumber markets. Modern devices for hauling logs, and improved sawing machinery are quite as important in the development of the lumber trade as inventions in wheat, cotton, or other industries. Bangor is one of the chief lumber centers of New England. St. Louis and Memphis receive and distribute much of the hard wood of the central region, and the Pacific coast ports are the natural markets for the lumber of the adjoining region.

106. Varying uses of wood. Notwithstanding the increased use of iron, stone, brick, and cement in building, it is found that the demand for wood is constantly increasing. For household furniture hard wood is the most suitable material and the demand is enormous. The same is true of agricultural machinery, railway cars, railroad crossties, all kinds of road vehicles, of ships, and of a vast number of tools and utensils used in shops, in the household, and on farms.

107. Wood pulp. The forests largely supply the modern demand for paper, the fibers of the softer woods being separated and made available by mechanical and chemical processes. From 1900 to 1907 the amount of wood pulp consumed in the

United States doubled, and the amount of wood used rose from 2,000,000 to 4,000,000 cords per year. Of all the wood employed in 1907 two thirds was spruce, followed by hemlock and poplar, with small amounts of pine, cottonwood, and balsam. Three Northern regions furnished most of the product, — namely, New England; New York and Pennsylvania; Michigan, Wisconsin, and Minnesota. Imports of spruce for this purpose are increasing, and nearly 1,000,000 cords were brought from Canada in

FIG. 83. Wasteful logging, Tyler County, Texas. The abandoned top would yield a timber 40 feet long and 10 inches square

1907. Some manufactured pulp also comes in from Canada and from Europe. Of 258 pulp mills in 1907, New England had 66 and New York 92.

108. Turpentine, rosin, and tar. These products are derived from certain species of pine. The earlier center of the industry was in North and South Carolina, but it has now passed to the southward, and Florida, Georgia, Alabama, Mississippi, and Louisiana are important producers. By tapping the trees resin is secured. From this by distillation turpentine is derived, and the residue is known in commerce and the arts as *rosin*. Tar is obtained by the direct distillation of the wood. The total annual value of these products in 1850 was less than $3,000,000, and in 1900 it was more than $20,000,000. Two thirds of the latter total was in spirits of turpentine, Georgia and Florida much surpassing other states. These products are exported to a considerable extent, and for this trade Savannah is the principal port.

109. Conservation of American forests. The first settlers in the virgin forest found the trees in their way, and their great

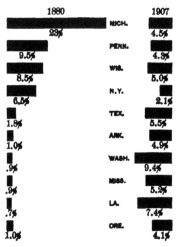

FIG. 84. Relative lumber production in ten states in 1880 and 1907

labor was to secure clearings for crops. The timber was burned, and, so far as needed, was made into charcoal. With the modern increase of population, wood has greatly increased in value, and high prices have led to wasteful cutting of the forests to secure the largest present gain, without reference to the permanent productiveness of the forest or the needs of coming generations. Forests are needed to promote even water flow, avert floods, prevent the destruction of the soil, promote river navigation, and provide for irrigation.

Scientific forestry as applied in Germany and other parts of Europe will secure permanent and regular crops of fuel and timber without injuring the forest. Only mature trees are cut, fires are prevented, and suitable planting and thinning receive attention. These methods are being applied in the United States, notably by the Forest Service of the Department of Agriculture, and by a similar bureau in the state of New York, where large forest reservations are owned by the state, in which it is hoped to develop the woodlands so that they may be permanent and at the same time make due return to the people.

The wood industries ranked fourth among the great groups in the census of 1900, and there is no hope of a lessened demand for

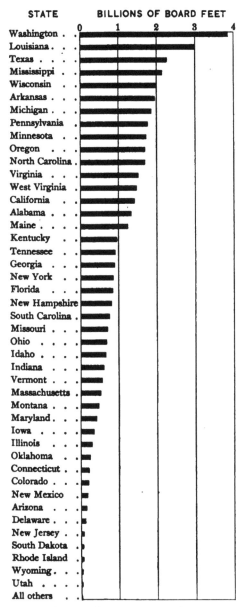

STATE — BILLIONS OF BOARD FEET

Washington
Louisiana
Texas
Mississippi
Wisconsin
Arkansas
Michigan
Pennsylvania
Minnesota
Oregon
North Carolina
Virginia
West Virginia
California
Alabama
Maine
Kentucky
Tennessee
Georgia
New York
Florida
New Hampshire
South Carolina
Missouri
Ohio
Idaho
Indiana
Vermont
Massachusetts
Montana
Maryland
Iowa
Illinois
Oklahoma
Connecticut
Colorado
New Mexico
Arizona
Delaware
New Jersey
South Dakota
Rhode Island
Wyoming
Utah
All others

FIG. 85. Lumber cut in 1907, by states

forest products. In view of the destruction that has already taken place, this is a serious problem, and has received its due share of attention in the recent discussion of the various conservation questions. It is held on expert authority that the Southern yellow pine will be exhausted in 18 years, and the Pacific coast timber in 41 years. Already the white pine of Northern forests has been brought to a low reserve. For industries and states which a shortage of hard woods would affect, the student may consult *Circular 116* of the Forest Service on "The Waning Hard-Wood Supply and the Appalachian Forests," 1907.

110. The use of the national forests. There were at the last report 145,000,000 acres of national forest lands, of which 5,000,000 were in Alaska and Porto Rico. The agricultural land included in these reservations is open to settlement, and mining is unrestricted. Timber is given away for domestic use and is

sold for commercial purposes. In 1906 there were given away 75,000,000 board feet, and 700,000,000 feet were sold. Grazing goes on as usual, but under restrictions which protect the rights of all. Fire protection is one of the chief ends sought, and the need of it may be seen in the fire-ravaged tracts of the Adirondacks, of southern Ontario, and of the Cordilleran highlands. It has recently been ordered that railways whose lines cross the Adirondack forests should, in the summer season, use oil as fuel for their locomotives. Progressive forestry clears away brush and fallen trees, thus lessening the danger of fire. Forest conservation insures a supply of timber for mining in the ore-bearing regions of the West, and the government forests become permanent playgrounds for all.

111. Foreign trade of the United States in forest products. The value of forest exports rose from $82,000,000 in 1904 to $104,000,000 in 1908. The large items are rosin, tar, and turpentine, sawed timber, boards, deals, and planks. A notable feature is the export, chiefly of Douglas fir, from Washington to Mexico, South America, and oriental countries; also consignments to Alaska and the Hawaiian Islands. From 1904 to 1908 Australia alone received from Washington and Oregon over 800,000,000 board feet of lumber, China and Japan over 700,000,000, and South America over 700,000,000.

Over against exports of $104,000,000 for the year ending June 30, 1908, there were in the same year imports of forest products to the amount of $101,000,000. The chief items were dyewoods and extracts; gums, such as gum arabic, camphor, and shellac; India rubber, which was much the largest item; cabinet woods and pulp wood. Mahogany came chiefly from Mexico and the Central American countries. Canada supplies most of the imported timber and pulp wood.

CHAPTER IX

ANIMAL INDUSTRIES OF THE UNITED STATES

112. General statement. Animals serve human needs chiefly in three ways, — in providing food, clothing, and power. While the more important are domesticated, wild creatures, game animals and fish, furnish considerable amounts of food. The cattle and dairy industries have been reviewed in Chapter III, and the other animal products will be found to illustrate the same general principles of commercial geography. Useful animals have their own climatic range, both as regards temperature and the vegetable products on which they depend. Thus the hog finds its most conspicuous place in the corn belt of the United States. Preservative processes are even more important than with most vegetable products, and the distribution of animal industry, as of all others, is controlled by the development of transportation systems which offer speed and reasonable rates.

The greatness of these industries is indicated by the Secretary of Agriculture in his report for 1907, in which he states that the live stock sold from farms plus that which is slaughtered on them is worth nearly twice as much as the cotton crop. In his report for 1908 he observes that the animal products amount to about three eighths of all the produce of the farm.

113. Sheep. These animals, like nearly all domesticated animals in North America, were imported from the Old World, where their origin from primitive wild creatures is lost in obscurity. The Spaniards brought sheep to Florida in 1565, and the missionaries took them to California in 1773. The colonies of Virginia and New York early received them from England and Holland. In 1810 interest in fine woolen goods led to the importation of 26,000 merino sheep from Spain. The merino and the English breeds are now of nearly equal importance in this country, the latter, or mutton sheep, being found in the East, where large populations require corresponding supplies of meat. Mutton, however, is not so general an article of food here as in Great Britain.

In 1900 more than half the sheep in the United States were found in the Western division. More than one fourth were in the states of the North Central division. The four leading regions were Montana, Wyoming, the territory of New Mexico,

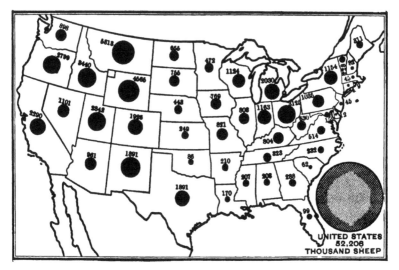

FIG. 86. Average annual number of sheep, in thousands, 1899 to 1908

and Ohio. In the Western division there were 33,000,000 sheep on 25,000 farms. In the North Central division there were 16,000,000 sheep on 358,000 farms. Thus great flocks are the rule in the West and on government lands. Because sheep often injure vegetation by too close cropping, grazing on government reservations is now placed under suitable restrictions.

114. Hogs. This type of domesticated animal has received its largest development in the United States, and it is claimed that American farmers have improved the breeds in a way similar to the advance made by the English agriculturists in beef cattle and in mutton sheep. Extremes of temperature are unfavorable for swine, and they require ample feeding. Thus the climate and the plant produce of the corn belt are favorable, and this industry has here become more important than elsewhere. It is also a natural accompaniment of the dairy industry, using the waste products of the latter. Water is essential, especially in summer; hence hogs are not much kept in the semiarid regions.

The meat and lard from the slaughter of hogs reported for the year 1906 amounted to more than 9,000,000,000 pounds.

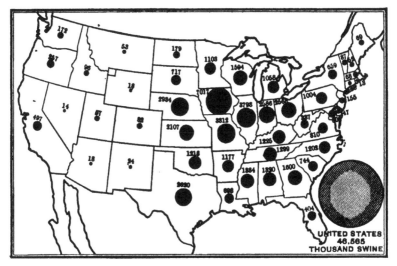

FIG. 87. Average annual number of swine, in thousands, 1899 to 1908

Pork affords the greater element in meat packing in the United States, — Kansas City, Omaha, and Chicago being the chief centers, with many other cities in the West and East. Cincinnati was formerly first in this industry, and earned the name of " Porkopolis "; but the migration of agricultural industries, due to the settlement of the prairies, carried the center of the pork industry farther west. The Southern states have moisture, mild temperature, and a plentiful supply of corn, and the industry is developing there. The growth of alfalfa under irrigation makes it possible also to raise hogs in the dry regions of the West.

115. The poultry and egg industry. The importance of this kind of farm produce is not well known because it enters little into foreign commerce and because the domestic trade is largely local. But here again the growth of cities and the facilities for rapid transit are enlarging the commercial scope of the product. Its importance is brought out by certain comparisons made by writers in the Department of Agriculture. Poultry produce is one of the four or five most important sources of agricultural wealth in the United States. It usually holds nearly a middle position in the list of eight or ten principal agricultural products of the several states. The egg product of the United States was more than the combined gold and silver output in every year from 1850 to 1899, though slightly exceeded by them in 1899. The poultry output also exceeded them in every year of that period except 1899 and 1900. Poultry and eggs formed, in 1899, one sixth of all the products of animal origin. At the census of 1900 there were, on the average, 42 fowls on every farm in the United States, the census not taking account of the large number kept in villages and towns.

Iowa, Ohio, Illinois, and Missouri were the four leading states, all belonging to the Central division. The growth in this industry in Kansas, another state of the same group, was from $6,000,000 in 1903 to $10,000,000 in 1907. The Secretary of Agriculture, in his report for 1907, says that the poultry products are worth more than the wheat. This emphasizes the difference between the two commodities. Both the foreign and the interstate trade in the one is vast, the market belongs to the world, and transportation and general finance are deeply concerned. The other trade is local, more widely diffused, is equal as a wealth producer, but less conspicuous. It holds a growing place, however, because of new attention to sanitation, to careful breeding, and to prompt marketing.

116. Wild game. This class of food was the principal support of the aboriginal Americans, and was an important part of the sustenance of the early settlers, but is now a luxury, often confined to the tables of the rich. The growth of population and the restriction of forests and other wild lands have put many species of animals in danger of extinction. Especially was this true when hunting for the wholesale trade became common. Hence legislation has become general in the states, restricting hunting to certain periods, usually protecting the various species through the breeding season, and often allowing hunting but a few weeks of the year. Some years ago the state of Connecticut made by law a closed period of eleven years for

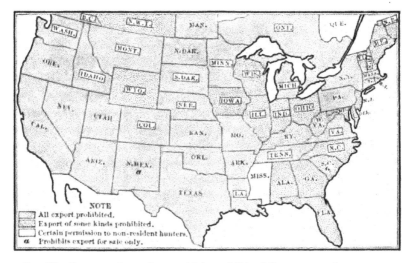

FIG. 88. States and provinces which prohibited the export of game, 1905

deer, with the incidental result that these animals became so plentiful as sometimes to injure the crops of the farmers. A beaver may not at any time be taken in the state of New York. It is well known that man has brought nearly to extinction the uncounted millions of buffalo that once trod the Western plains, and now the governments of Canada and the United States are guarding the few which remain, by caring for them and seeking to increase their numbers on government reservations.

The federal government has begun also to protect other wild life, the first general law having been enacted in 1900. This law provided for information, for regulation of importation of foreign species, and for the restriction of interstate commerce.

Much traffic in game, especially birds, centering in Chicago, St. Louis, and other cities, was stopped, preventing the complete destruction of grouse, prairie chickens, quail, and ducks. The wholesale game trade thus passed out of existence. The general government has coöperated with the states and with the Audubon societies everywhere, resulting in discontinuing the use

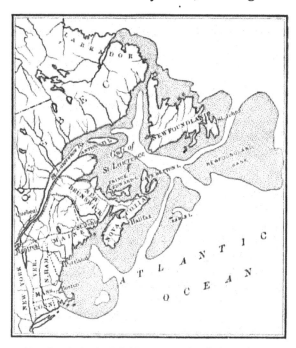

of plumage of native birds for millinery, and in stopping the export of native cage birds. Some game preserves have lately been set apart, usually small areas serving as breeding grounds for birds. An example is Pelican Island, in Indian River, Florida,—a tract of four acres, a breeding place for the brown pelicans. Breton Island is a breeding

FIG. 89. Principal fishing grounds of the Atlantic coast region

place of terns, off the mouth of the Mississippi River ; and the Wichita preserve of 57,000 acres has been set apart in Oklahoma for birds and big game.

Man's power to change the life of the earth is nowhere better seen than in this field. With growing population, effective weapons of destruction, and modern transportation, he would, unless restrained by public sentiment and by law, destroy the desirable species through greed, and, by carelessness or ignorance, spread widely the undesirable species.

117. Fisheries. Under this head other aquatic foods as well as fish are included. The waters of the world afford a vast field for the growth of food materials, although it is the shallow marginal seas which yield the greater abundance, and it is these over which man may hold more effective control. The product belongs more to temperate and northern seas, and thus, like products of the soil, is somewhat subject to the larger elements of geographic influence.

The fishing regions of the United States may be given a fourfold division.

1. *The Atlantic coast fisheries.* These are most extensive along the New England shores, where the continental shelf and the tidal bays offer favorable conditions for fish. The cod, herring, and mackerel are the chief species taken, and the cod is chiefly sought on the Banks of Newfoundland, a wide area of shallow water extending from that island far into the Atlantic. This is a most historic fishing ground, frequented by English, French, and Canadian fishermen, and by those from the United States. Of late the industry has become much centralized, being largely confined to Gloucester and Boston. Within the " three-mile limit " fishing belongs exclusively to the adjacent country, but, outside of that, fishing grounds are recognized as the common property of nations.

2. *The Pacific coast fisheries.* In this region the salmon is the chief fish of commerce. When the first explorers crossed the mountains they found the red men living chiefly upon it, and its capture, canning, transportation, and sale have long afforded an important total among the animal products of the Pacific coast. The salmon abounds in the waters of the Columbia and Fraser rivers. Astoria, at the mouth of the Columbia, once important as a depot for furs, has now become a center of the salmon industry. These fish periodically enter these streams and follow them hundreds of miles for spawning. They are captured, preserved both by cooking and hermetic sealing, and sent by means of refrigeration and quick carriage to eastern markets. The Alaskan salmon fishery has grown to large proportions.

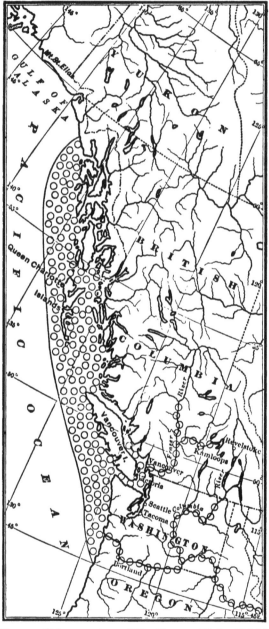

FIG. 90. Salmon waters of Washington, Idaho, British
Columbia, and Alaska

3. *Great Lake fishing*. The Great Lakes afford an important supply of some kinds of fish, particularly of lake trout, white-fish, and sturgeon. The development of fresh-water fish-eries, with rapid-transit facilities, is believed to be one of the factors in the decline of the New England fishing interests. The sturgeon is also to be found in marine bays, as the Chesapeake and the Delaware, and while little in demand until about fifty years ago, in recent years it has been especially desired for its roe, so that the fish has been sought in the breeding season, and hunted down till it is well-nigh exterminated.

4. *The smaller lakes*, and inland

streams. A very considerable volume of food is gained from these widely distributed fishing waters, but little of it enters into the commerce of the country.

118. Other aquatic foods. Here belong oysters and other so-called shellfish, — lobsters, crabs, and shrimps. Of these in American waters the oyster is chief, and is especially abundant

FIG. 91. Oyster bed at low tide near Brunswick, Georgia

on the Atlantic coast from New England to Chesapeake Bay. In many beds young oysters are " planted " until they mature, and are then systematically gathered for market, — a process resembling the farmer's intensive culture of his fields. Oysters are sent inland chiefly by refrigeration, but sometimes as a canned product, and a considerable quantity is exported. The oyster product of the world, outside of the United States, is insignificant.

119. Organization and protection. The United States Fish Commission is an important organization, employing experts for the dissemination of knowledge, supporting laboratories, and coöperating with state institutions. State laws protect fish as they do wild game, and, by means of hatcheries, lakes and streams are periodically stocked with the young of desirable food fishes.

120. Bees and their products. Bees furnish another illustration of man's power to change the grouping and distribution of animals and plants throughout the world. Honeybees were brought from the Old World to the New in the days of the

FIG. 92. Gloucester harbor and fishing vessels

early settlements, and at various times in the past fifty years the Department of Agriculture has introduced important varieties. In the Northern states the bees gather the nectar largely from clovers, buckwheat, the linden, and the raspberry. The tulip, palmetto, gum, mangrove, and citrus are sought by the bees in the South. The wax is an animal product more truly than the honey, for it is secreted by certain glands belonging to the body of the bee. In 1899 the five leading states in producing honey were, in order, Texas, California, New York, Missouri, and Illinois. Beekeeping is more associated with fruit culture and dairying than with other kinds of farming. The total production of honey in the

United States in 1899 was 61,000,000 pounds, of which more than half came from the North and South Central divisions.

121. Wool. In 1850 the North Atlantic and North Central States produced most of the wool grown in the United States. That grown in the more westerly region was insignificant. In 1900 much more than half the total was produced in the Western division, and somewhat more than one fourth was grown in the North Central States. Thus the industry has moved westward. From 1850 to 1900 the number of sheep increased 67 per cent and the wool clip 300 per cent. Better breeds and better care resulted in much heavier fleeces. The wool clip of the United States for 1908 was about 311,000,000 pounds. The total domestic supply falls from 25 to 50 per cent below the demand, the deficiency being supplied from abroad. The temperate lands of the southern hemisphere supply the needs of the manufacturing countries. Their wide and thinly peopled plains and mild winter climate furnish favorable conditions. Here belong Argentina, South Africa, and Australia. Boston is the most important of American wool markets.

122. Leather. The geography of leather making depends upon the supply of hides and skins, upon suitable conditions for tanning, and upon the location of markets. Upon all these, however, modern transit and the invention of advanced processes have an important influence.

Tanning was early established in the Massachusetts and Virginia colonies; the former long held the leading position, but at the census of 1900 had dropped to fourth place among the states, Pennsylvania ranking first, New York second, and Wisconsin third. Tanning has followed the supply of oak and hemlock bark, materials which are still most largely used. The " chrome process " has, however, taken a considerable place in making some products, especially upper leather. Modern invention utilizes the entire tannin contents of the bark and brings results more swiftly than the methods of the early days.

The hides and skins are drawn from many sources. The leather industry is an accompaniment of live-stock raising, utilizing the skins of cattle, horses, sheep, pigs, and goats. To a lesser degree the skins of some wild creatures enter into the making of leather. Among these are deer, kangaroo, porpoise, seal, and alligator. American leather makers import skins from many lands. Many come from the European ports, — London, Hamburg, Marseilles, Naples, Salonica, and Constantinople; also from India, China, and several South American countries. The capital employed in the industry has steadily increased in recent years, while at the same time the number of tanning plants has decreased, thus showing the progress of consolidation and the elimination of small concerns.

123. Furs and feathers. Fur-bearing animals are in great part the inhabitants of the colder regions, and hence furs are largely the product of arctic and subarctic lands. As the cold lands are chiefly in the northern hemisphere, this commodity largely originates in that part of the world, but it is manufactured and used chiefly in the north temperate belt, where the winters make such protection useful, and where live the greater number of civilized people. Canada and the Russian Empire are the principal fur-yielding territories. The United States has its most immediate interest in the Alaskan seal fisheries. Several minor fur-bearing animals are still found in the United States, such as the raccoon, mink, muskrat, skunk, and fox. Before the wild lands were so fully subdued the United States produced more furs than now, sharing in that early trade in beaver and other peltry which furnished a most important motive in early explorations, and which formed an object of trade jealously guarded by French, Dutch, and English colonies. The trade is more concentrated in New York City than elsewhere in this country, for here are received both undressed and manufactured furs from London, Leipzig, and other foreign marts, as well as domestic supplies.

With the suppression of the use of the plumage of native wild birds, ornamental feathers are little produced in this country. An interesting exception is found in successful beginnings in ostrich farming in Arizona, California, Florida, and Arkansas.

124. Silk. Practically all the raw silk of the world is produced in Europe and Asia. The United States, through the Department of Agriculture, is carrying on experiments in this direction, but the results have not yet become commercially important. Although all the raw material is imported, the United States now manufactures more silk than any other country. The silkworm, and the mulberry upon which it lives, are suited to other continents than those named, but require cheap labor and skill in handling, neither of which conditions has as yet developed in other parts of the world. Wool, therefore, and leather are the principal clothing materials of animal origin grown in the United States.

125. Animals used for power. Of these the most important is the horse, cattle being little employed for this purpose in the United States. About 21,000,000 horses were enumerated by the Twelfth Census. The Western states raise more horses for market than other parts of the country. Most of the horses for city use and for export are reared west of the Appalachians, Omaha and Kansas City being the more important markets. More people own horses in small than in large cities, because in the latter other means of conveyance are abundant, and because horses are less easily kept in such centers. The fears that bicycles and automobiles would drive out horses have not been realized, perhaps because the farms require 80 per cent of all the horses bred each year. Since agriculture is growing and the export demand is increasing, the future of the horse seems assured. As the Erie Canal and the railways, contrary to the expectation of some, increased the demand for draft animals, the present development in mechanical transportation will not be likely to reverse past experience.

126. Organization in animal industry. The Bureau of Animal Industry in the Department of Agriculture has done much in this direction. It disseminates information, promotes the introduction and breeding of new strains, studies the best methods of feeding and care, and, by instruction and rigid supervision, prevents disease, and suppresses it when introduced. When necessary, local, interstate, or international traffic is regulated or forbidden, and infected herds are slaughtered, with liberal compensation to owners. General authority in these matters must belong to the central government, as state authority would often be powerless.

State bureaus, agricultural colleges, and experiment stations act in coöperation with the general government. Acting for the same ends — of better breeding, care, and sanitation — are

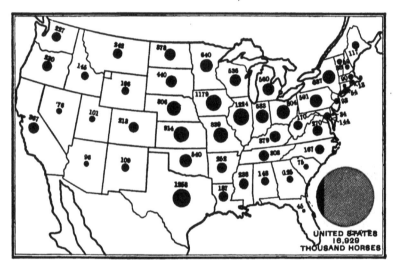

FIG. 93. Horses on farms and ranges, average annual number in thousands, 1899–1908

numerous state and private stock breeders' associations. Some of the latter are national in scope, giving information through literature and through fairs and other public exhibitions, and using influence for favorable laws concerning the stock industries. Such associations pay much attention to the breeding of blooded stock, showing its advantages and protecting their members against dishonesty and false pedigrees. Many associations in the range country protect against thieving and false branding. Sheep breeders often organize against the encroachments of cattlemen.

127. Miscellaneous animals and animal products. As has been already stated in reference to cattle, the bones, horns, hoofs, and other parts of all domestic animals are used for the making of notions and various objects, and the final refuse is an important contribution to fertilizer. Soils are not ·inexhaustible, and wisdom requires that all organic material removed from them by the processes of growth should, so far as possible, be returned to them. The bones and oil of the whale offer other

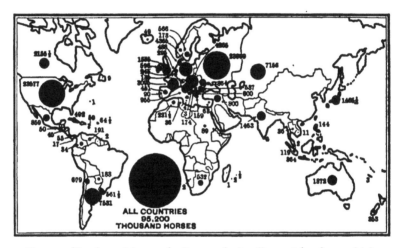

FIG. 94. Number of horses in thousands, in all countries from which data are obtainable

examples of animal products. Whaling, once an important marine industry of the United States, has declined because the supply is small and because other substances have taken the place of both the bone and the oil. Sponge fisheries exist on the Florida coast in the neighborhood of Key West.

CHAPTER X

MINERAL INDUSTRIES OF THE UNITED STATES

128. Distribution of mineral and organic products. Minerals have little present relation, as regards distribution, to climate or to conditions that make climate, such as latitude, altitude, rainfall, temperature, or the movements of the atmosphere. There have been strong relations in the past, as in the climate of the coal period. So far as peat is forming to-day, climate is influential in determining its presence. Coal and peat are, however, primarily organic, although usually classed among minerals. Temperatures and the presence of water have in the past had much to do with the making of some mineral deposits, but now we may find gold in the arctic belt in Alaska, in the temperate regions of the Cordilleras, and in the tropical belt in Mexico. Even more widely distributed are building stones and clays.

129. Causes of the distribution of mineral products. These are various, and their full study belongs to geology. One broad distinction, however, may be made in a study of commercial geography. Mountain lands, because regions of disturbance, have been favorable to the making of some mineral deposits, especially of the metals. In such regions heat, water, and pressure have worked effectively in separating the minerals from the rocks and in forming workable ores. Thus gold and silver abound in the states of the Western division. The rule has many exceptions, however, for the precious metals are not important in the Appalachians and they scarcely occur in the Alps. Lead and zinc occur most largely in regions of disturbance, such as the Ozark plateau and the Rocky Mountains, but are also found in the undisturbed rocks of the prairies of Illinois and Wisconsin.

Petroleum and natural gas are not found in regions of disturbance, for, although they may have occurred before the rocks were

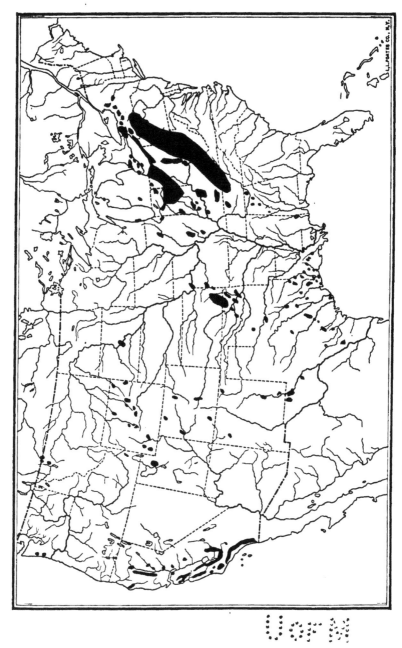

FIG. 95. Petroleum and natural-gas fields of the United States

upturned, the cracking and thrusting on edge have released these light and often gaseous products, so that they have long ago been dissipated in the atmosphere or have been swept down the streams.

Building stones occur in both kinds of territory, but their character varies. Thus granites, marbles, and slates are found in regions of disturbance, while sandstones and ordinary limestones often occur elsewhere. Rocks of volcanic origin depend upon the outflow or outburst of heated materials, and may belong to regions of present or of ancient volcanic activity.

130. Some minerals exhaustible. A definite quantity of some minerals exists in workable masses, and this amount is enough for the uses of man for but a limited time. Geological periods have been required for their formation; and if they are still making, the process is so slow that the supplies thus formed will not be of practical interest to the men of to-day. Such minerals are gold, silver, coal, copper, iron, and petroleum. If we use them all, there will for thousands of generations be no more.

The case is otherwise with the products of the soil, which, with reasonable care and wise management, may be won indefinitely. Building stones, also clays, cements and cement materials, sands, gravels, and salt occur in such quantities as to be practically inexhaustible.

131. Petroleum. The mineral oil bearing this name began to be important as the foundation of an industry in 1859, and Pennsylvania was practically the only producing state until 1875. In those years lighting oil derived from petroleum largely replaced animal substances such as whale oil and the tallow candle. It is now to a considerable degree replaced by electricity, and thus lighting offers an example of the importance of invention and discovery in serving the convenience and changing the industries of civilized peoples. Petroleum products are now much used for the purpose of lubrication, and as they are less expensive than animal oils there is no apparent substitute for them.

The principal oil fields in the United States are here indicated: the Appalachian field occupies the same general region as Appalachian coal; oil is not, however, found in the anthracite region,

·but it occurs in western Pennsylvania, along the Allegheny rather than along the Monongahela River. It is found extensively in Ohio and West Virginia and as far south as Tennessee. Indiana also produces petroleum, and there are fields in Kansas, Oklahoma, Texas, Colorado, Wyoming, and California. The Texas field is along the Gulf coast and extends into Louisiana, the most important center being at Beaumont, where a vast number of wells have been sunk on a small tract of 200 acres. California has several fields, all in the southern half of the state, and has far outstripped any other state in production in recent years. This abundance is highly important because that part of the Pacific region has so little coal, and indeed imports much of this fuel.

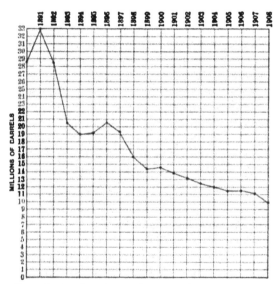

FIG. 96. Decline in production of New York and Pennsylvania oil fields

The highest point in Pennsylvania production was reached about 1890, the output now having dropped to one third, as compared with that time. The amount stored in the rocks is limited and is rapidly exhausted. The so-called "pools" are not cavernous reservoirs, but occur in masses of porous rock, usually sandstone, — the "oil sands" of the well driller. Sometimes the oil is forced out by a vast pressure of natural gas, but more often it is pumped, at the rate of a few barrels per day. The life of a well is prolonged by " shooting " it, that is, by exploding a heavy charge of nitroglycerin at the bottom, shattering the rock and allowing the oil to concentrate. Wells vary in their period of production from a few months to a score of years or more. Seven years is considered in Pennsylvania to mark the average life of a well. The output has also declined in Ohio, West Virginia, Kansas, and Texas. Experts hold that the present known supply of petroleum in the fields of the United States, at the present rate of production, will be exhausted in ninety years. If, as in the past, more is used each year, the time will be shorter. The life of the industry may be prolonged by fresh discoveries.

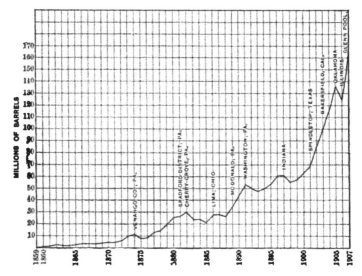

FIG. 97. Annual production of petroleum in the United States, 1859–1907

The transportation of petroleum has led to the adoption of ingenious methods. As it is inflammable, iron reservoirs and tank cars are employed, but a large part of the crude product is conveyed in pipe lines to remote cities. Smaller pipes collect the oil in a given field, and it is then forced, or flows by gravity, for hundreds of miles; as from western Pennsylvania to Baltimore, Philadelphia, New York; or to Buffalo, Cleveland, and Chicago on the Great Lakes. The small reservoir at the well, the small pipe, the large reservoir, the great pipe, the tank steamer, the reservoirs in remote seaports, and tank cars in distant lands form links in the chain of transportation by which

American oil is transferred from Pennsylvania or Ohio to light homes on the other side of the world.

Refineries are established in the oil fields, at the ports as above named, and some American petroleum is refined in foreign lands. A great number of products results from the processes of refining. Among these are gasoline, benzine, lighting oils, lubricating oils, vaseline, and paraffin. Russia, with the fields of Baku on the Caspian Sea, is the only great rival of the United States in the petroleum industry.

132. Natural gas. This substance is often found in petroleum territory, but may occur alone, as in several places in central and western New York. The largest fields are in western Pennsylvania and northern Indiana, but the size of the field does not indicate the amount of gas now available. Indiana reached its highest production in 1900, and in 1907 had declined more than three fourths. In the latter year the leading states were, in their order, Pennsylvania, West Virginia, Ohio, and Kansas. The total product for the United States in that year was nearly $53,000,000, the highest value from the beginning of the industry. Natural gas is now transported largely from West Virginia to Cleveland and other Lake Erie ports. This shows the possibility of piping ordinary fuel gas long distances, and introduces an important factor into the development and localization of industry.

The pressure of gas in new wells is sometimes very great, amounting in one case (New York, 1897) to 1500 pounds per square inch. No well is likely to show a pressure beyond 100 pounds to the inch after 10 years, and the yield is prolonged in an important way by pumping the gas. The waste of gas is large, due sometimes to the difficulty of capping wells subject to great pressure. On high authority it is believed that even now 1,000,000,000 feet of gas are wasted every 24 hours. This waste is an important item in the general movement for saving the natural resources of the United States.

133. Bituminous substances. These, as found in the United States, are divided by Ries into two groups : the asphaltites, or masses of purer solid bitumens; and the bituminous rocks, whose pores are filled with this material. In either case it is believed to have been derived from petroleum. The most important deposit of asphaltite in the United States is in Utah. The bituminous rocks are found chiefly in Kentucky, Oklahoma, and near Santa Cruz in California. The chief supplies, as for paving, are, however, drawn from Trinidad. Some forms of the bitumen serve minor purposes, as varnish for ironwork, roofing pitch, and insulation of electric wires, as substitute for beeswax, and as a constituent of candles.

134. Salt. This substance was early obtained in the Atlantic colonies by the evaporation of sea water. Sometimes the boiling process was used and occasionally the sun's heat was employed. This industry continued in Massachusetts until after 1830, when the salt from Onondaga County, New York, became so abundant and cheap as to absorb the market. Salt was made about San Francisco bay from the time of the discovery of gold in 1849. There the industry continues, and has the advantage of prolonged dry seasons, so that the vats need not be covered from storms, as in the East. The wind also, being constant, favors evaporation, and as it comes from the Pacific Ocean it does not bring dust to the salt. The Mormons gathered salt on the shores of Great Salt Lake, beginning with 1847. Michigan, in the Saginaw region, began to be an important producer in 1859, and Kansas in 1867. New York and Michigan have been the leading states, each having in some years been the largest producer. Other important states are California, Kansas, Ohio, Texas, Utah, and Louisiana. The eight states named produced 98 per cent of the salt output of the United States in 1905. The total in that year was a little over 17,000,000 barrels.

Salt is derived as rock salt, being mined in its natural state as associated with beds of rock; also as solar salt, evaporated by the sun's heat; and by boiling processes. It is further found as a natural surface deposit in arid regions, especially in connection with salt lakes. The rock salt is due to natural evaporation in more or less remote geological times, and was afterwards covered by rock deposits. Underground waters in some places meet these beds and form brines, which are reached by wells, or come to the surface as salt springs. The latter, known as " salt licks," were common in Kentucky and elsewhere, and were frequented by wild beasts and by aboriginal tribes. Thus the Indians long made salt at the Onondaga Springs in New York. White men began to make salt there soon after the Revolution. Only brines were found, first in springs and later by boring. Salt formed the

FIG. 98. Salt works, shore of Cayuga Lake, near Ithaca, New York. The vats
for the brine are above the building on the right

fundamental and leading industry of Syracuse until the time of the Civil War, when the Michigan product and foreign salt came into competition. In 1880 beds of rock salt were discovered near the Genesee River, and now the salt industry of New York has chiefly moved westward from Syracuse, the salt being obtained by dissolving and evaporating and by direct mining processes.

The largest use of salt is for seasoning and preserving food. More than three fifths of the domestic product in 1905 was so used. Above one fifth was used in making soda products and bleaching materials. Smaller quantities were consumed in making various chemicals, in the curing of hides, in freezing mixtures, and in the manufacture of pottery, dyestuffs, fertilizers, fire-clay products, brick, and tile.

135. Fertilizers. The chief mineral materials so used are gypsum, phosphates, limestone, and marl. Gypsum is sulphate of lime and is often found in beds in connection with rock salt. New York, Michigan, Kansas, and Iowa supply most of the gypsum used in the United States. Deposits occur southward from Kansas, in Oklahoma and Texas. The substance has other uses, as in the manufacture of plaster of Paris, and when used as a fertilizer is often called "land plaster."

Phosphates have been produced in South Carolina since 1867, but Florida has taken first place for several years. Phosphates are also found in Tennessee, Arkansas, and in an area extending from Utah into adjoining corners of Wyoming and Idaho, a region said to "embrace the largest area of known phosphate beds in the world." Of 2,265,000 tons produced in 1907, exports made 40 per cent. This seems to be an unfortunate item of foreign trade, in view of the vast areas of agricultural land at home, which need, or will soon need, the application of fertilizers.

Other mineral substances used in a limited way as fertilizers are quicklime, made from limestone; marls, consisting of varying mixtures of clay and lime; and the so-called greensand, containing iron, potash, and phosphoric acid, and occurring chiefly in the young strata of the coastal plains of New Jersey and Virginia.

136. Building and ornamental stones. The mineral fuels are found in relatively few parts of the world, where ancient conditions have favored their formation. The same is true of salt and of mineral fertilizers. Building stones, on the other hand, are widely distributed, because rock-making agencies have been active almost everywhere. Some ancient rocks, indeed, are so soft or so full of joints that they are unsuitable, and along the borders of seas the rocks are sometimes too young to have acquired the necessary hardness and strength; but it is still true that a supply of suitable stone is usually found near the place where it is desired to use it.

Because most building stones are abundant and of small value in proportion to their weight, the trade in them is local. Only a few enter into international or even into interstate commerce. Such heavy freight is transported more readily by water than by land; yet building stones are so commonly found in inland situations, or among rugged hills and mountains, that the latter mode of carriage is much more common. Here is found a mineral commodity, which, unlike petroleum, iron, coal, or gold, is, for human needs, absolutely inexhaustible.

The use of stone depends partly upon the abundance of the forests, and hence is apt to be larger in the older lands, in which the supply of wood is restricted. Also, with growing wealth and mature civilization men desire to erect monuments and buildings of the highest degree of beauty and permanence. Many classic structures of the Old World have been the product of the religious and artistic aspirations of the people, aided by an abundance of cheap labor. The United States is a new country, with its structures chiefly of wood, both because wood has until recent years been plentiful and cheap, and because a people absorbed in subduing a continent could not, in general, erect the more durable structures. The tendency is now toward the use of stone, especially in cities. For foundations stone is commonly used, as well as in bridges, embankments, locks, and other works of a private or public character.

It will be enough to name five of the more common kinds of stone that meet these needs; namely granite, slate, sandstone, marble, and ordinary limestone. Granite is a term properly used of a crystalline rock consisting largely of three minerals, — quartz, feldspar, and mica. The two first named are hard minerals, and aggregates of the three, especially where fine-grained, offer the qualities of resistance, hardness, color, and durability, which are important in building material. For the student of commercial geography the distribution of the granites is the most important fact concerning them. Granites are usually among the older rocks, and often form the central or axial parts of mountain ranges. Four chief regions may be distinguished in the United States : the Appalachians, from Maine to Alabama; the Lake Superior region in Wisconsin and Minnesota;

the Rocky Mountain belt; and the Southwest, along the Sierras and running through southwestern Arizona. The most extensive quarries are in the northeast, because the population is dense and much granite is required for city buildings and burial monuments. Westerly, Rhode Island, Quincy and Rockport, Massachusetts, and Barre, Vermont, are among the best known quarrying centers for this kind of stone.

Slates are rocks which cleave into thin sheets, and, being hard and resistant to weathering, are useful for roofing and other purposes. They originate in a soft mud rock or shale, which has been changed under pressure to the present condition. The process of change is known in geology as metamorphism, and consequently slates are found only in regions of disturbance, where the necessary forces have been available. The chief region in the United States is along the Appalachian belt, but slates occur in the Lake Superior region, in Arkansas, along the Sierras, and elsewhere. The Appalachian slates enter most into trade because the dense populations are in the East. Extensive quarries are worked along the New York-Vermont border; also in Maine, Pennsylvania, and southward to Georgia.

Sandstones, being cemented masses of sand, have been formed in all ages and are widely distributed. But few in the United States are used or transported widely enough to warrant a place here. The " brownstones," which indeed may be brown or red, once extensively used for city houses, come from the Connecticut valley in Massachusetts and Connecticut, also from New Jersey and Pennsylvania. The Potsdam of New York is a very hard red sandstone, and the Medina of the same state is red and gray and much used for paving. Fine-grained sandstones that come out of the quarry in wide, thin slabs are known as flagstones and are used for sidewalks and curbs.

The Berea sandstone of Ohio now enters into interstate commerce more widely than any other.

Limestones are chiefly of organic origin on the floor of ancient seas, and are abundant, though less so than sandstones. They consist mainly of carbonate of lime, or of lime and magnesia, and are a rather soft and soluble rock, though the more compact varieties offer good grades of stone. The Bedford limestone of Indiana, consisting of small rounded or oölitic grains, is widely used, and much is shipped to Eastern cities. It is readily dressed to any desired form, is of good color, and is reasonably strong and durable.

Marbles are metamorphic limestones, and hence, like slates, are confined to regions of disturbance. In the language of trade,

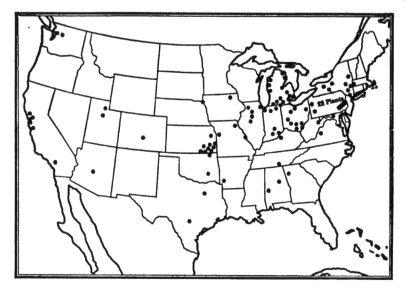

FIG. 99. Portland cement plants in the United States. (After Eckel)

any limestone that will take a polish is likely to be called a marble. True marbles occur most extensively and have been most developed along the Appalachians, the largest American quarries being at West Rutland and Proctor, Vermont. Massachusetts, New York, and Tennessee furnish other Eastern marbles, and Colorado and California offer a future supply for the West.

Ornamental stones, including gems, are not so abundant in the United States as to require an account in an elementary work. In 1907 the stone product of the United States, including furnace flux and lime, amounted to about $90,000,000.

137. Cements. These materials are derived largely or wholly from stone, are used for similar purposes, and are of growing importance in the United States. Ordinary mortar is chiefly a compound of sand and quicklime, which has the property of setting and thus becoming like rock. Certain impure limestones, when burned and ground, form a natural cement, which has hydraulic properties and will set under water. Among the chief localities in the United States are Rosendale and Akron, New York, and also Cumberland, Maryland. Much more extensively used at the present time are the Portland cements. These are made by burning certain combinations of limestone, marl, and clay. The kiln is kept at a high temperature and the resulting clinker is ground. The powder

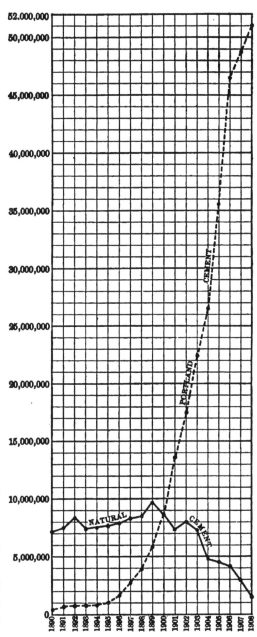

FIG. 100. Comparison of production of Portland and natural cement, in millions of barrels, 1890–1908

sets under water like natural cement, and is also widely used in constructions above ground. This industry was of small extent as recently as 1890, but has now grown to large proportions. In 1900 the total production in this country was valued at over $9,000,000, in 1903 at nearly $28,000,000, and in 1908 at over $40,000,000. The supply of the necessary materials is practically without limit, and the growth of population and the increasing cost of wood indicate a vast future use of this material. New York, New Jersey, Pennsylvania, Michigan, and Ohio are among the heaviest producers, a large percentage of the total being made in the Lehigh district of Pennsylvania.

138. Clays. Like the common rocks, clays are widely distributed and are used for so many important purposes that they enter largely into manufacture and commerce. They consist of the finer waste of the surface rocks, the essential element being aluminum silicate. They are derived from the rocks by weathering and mechanical erosion, and are deposited by streams and in lakes and seas. Where lakes have been partly or wholly drained, or the borders of seas have receded through uplift of the land, areas of clay are often exposed and made available. Ancient marine clays, long hardened into rock, are often quarried and ground, thus offering supplies of clay where there are no surface beds of the unconsolidated material.

Clays lend themselves to many uses, as for building brick, fire brick, sewer pipe, drain tile, pottery, and paint, and as a constituent of some papers. Kaolin is a pure clay produced by the wasting of ancient crystalline rocks. It is worked chiefly in the eastern states, notably Pennsylvania and North Carolina, and is employed for pottery. Lower grades of clay are used for stoneware. Brick clays are found in all states, but noteworthy development has occurred where abundant deposits are available by water carriage to great cities. The most conspicuous illustration is found along the Hudson River, where large deposits were made in lake waters at the close of glacial time. These clays are made into bricks and floated in scows to New York City. The brick clays in the Atlantic States from New Jersey southward are mainly marine, being found in strata of the coastal plain. In the states between the Ohio River and the Great Lakes alluvial clays and glacial-lake clays are the common materials for brick.

The total clay product of the United States in 1905 was $135,000,000, and Ohio, Pennsylvania, New Jersey, Illinois, and New York, in the order named, were the leading states. Bricks are required in increasing quantities in paving the streets of cities and towns; and according to recent reports of the census, great fires, as at Baltimore, have increased the demand for hollow building blocks and other noncombustible products of clay.

139. Various mineral products. Graphite is a black mineral composed chiefly of carbon, and is often called black lead. Its most familiar use is for lead pencils, but it is also employed in making crucibles, lubricating powder, stove polish, and for other purposes. The chief locality in the United States is Essex County, New York, near Lake Champlain, but the larger reliance is on importations from Ceylon.

Abrasives are used for various sorts of grinding and are found in various forms. Some are hard rocks, suitable for grindstones, millstones, and whetstones. The Berea grit, an Ohio stone, is chiefly used in the United States for grindstones. A very fine-grained sandstone, called novaculite, is quarried in Arkansas, and is used at home and is exported, for hones and oilstones. Some hard minerals are crushed or powdered, and are used in the manufacture of sandpaper and abrasive wheels. Quartz, garnet, and corundum are such minerals. Corundum mixed with certain kinds of iron forms emery, of which the chief deposits are at Chester in western Massachusetts. Various abrasives are produced by artificial processes and are much used.

Glass is made from quartz sand which is widely distributed over the United States. The value of the glass product in 1905 was $79,000,000, having risen to that figure from $21,-000,000 in the previous twenty-five years. Of the total, building glass made a little more than one fourth, pressed and blown glass about the same amount, while bottles and jars aggregated $33,000,000. Pennsylvania stood first, with $27,000,000, or a little more than one third of the total. Indiana came next, with $14,000,000, followed by Ohio, New Jersey, Illinois, and New York. Of all the building glass made in the United States, Pennsylvania produces more than one half. The location of glass factories is influenced more by the availability of suitable fuel than of the sands. It is stated that good glass can be made from impure materials with good fuel. A glass factory was built in Virginia in 1608, and the industry remained on the Atlantic border until 1797, when coal was adopted as fuel by an establishment in Pittsburg, in a state which has continued to provide for more than a century the best of fuel for this industry. Natural gas is one of the best fuels, but is less permanently available than coal, while oil is much used but is considered expensive. "The cost of fuel is the largest item of expense in glassmaking" (United States census report). In recent years important machinery has been adopted in glassmaking, and importations from France and Germany have greatly declined.

140. Metallic and nonmetallic minerals. All the mineral substances thus far studied in this chapter are known as nonmetallic. As regards their importance to man, a threefold division may be made : first, salt ; second, the group of fuels, including coal (Chapter V), petroleum, and natural gas ; third, the building materials, — stones, cements, and clays.

Of the metallic minerals iron has already been considered (Chapter IV), and is by far the most useful to man. The precious metals have important uses as money and in the arts, but the absence of them would not be fatal to commerce and the general progress of the race.

141. Gold. According to figures given by the United States Geological Survey, $60,000,000 per year in gold was the amount of the output soon after the gold discoveries of California in 1849. By 1883 production decreased to $30,000,000. A few years later the region about Cripple Creek, near Pikes Peak in Colorado, opened, and production reached $80,000,000 in 1902.

Alaskan discoveries followed, and the output in 1906 reached $94,000,000. The world's production in 1907 was $412,000,-000, of which $151,000,000 came from Africa, mainly from the Transvaal. The most recent important discovery in the United States is in Nevada. Three states and one territory have now been named. Other chief producers are South Dakota,

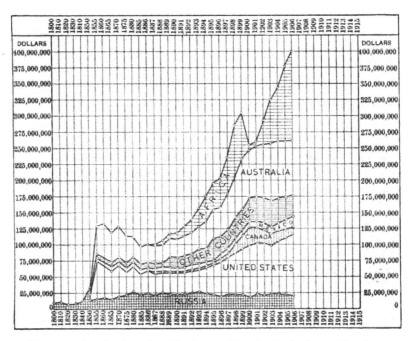

FIG. 101. Production of gold; the world and principal countries from 1800 to 1906

Montana, Idaho, Oregon, and Arizona, so that nearly all the gold of this country comes from Cordilleran territory. As gold offers great value in small compass and is easily transported, the limitation of distribution is not unfortunate. It would be far otherwise if such essential products as coal, iron, or stone were confined to a single region, since they are both bulky and heavy and therefore costly to transport. California and Alaska gold is found in placers, that is, in a fine fragmental state among gravels of present or ancient streams and seas. This gold is secured by "washing" the gravels, and the process is known as hydraulic mining. The chief gold supplies of other states are won by underground mining, following the course of gold-bearing veins. The experts of the Geological Survey express the opinion that the gold output of the United States is not likely in coming years to rise above $60,000,000 or sink below $40,000,000 per year. In 1906 about two thirds of the gold used in the United States was coined and one third was employed in the arts.

142. Silver. This metal was not produced in an important amount in the United States until about 1860. Soon followed great discoveries at Virginia City in Nevada; also in Arizona, Montana, and at Leadville and other centers in Colorado. The maximum was 63,500,000 ounces in 1892. The price, however, declined from $1.24 per ounce in 1875, to 53 cents in 1902, rising to 67 cents in 1906. The United States and Mexico, in about equal measure, contribute approximately two thirds of the world's supply. Canadian production is increasing, reaching nearly 20,000,000 ounces in 1907. The rank of states in 1907 was as follows: Colorado, Montana, and Utah were in a first group, producing about equally; then came Nevada and Idaho, followed by Arizona and California. Thus the silver in this country is chiefly Cordilleran. Much of it is really a by-product secured in the mining of lead and copper. More than half the product is exported. Relatively more silver is used in the arts than is the case with gold.

143. Copper. In 1908 the United States produced 943,000,-000 pounds of copper, which was more than four times the output of 1888, or 226,000,000 pounds. This in turn quadrupled the production of 1878, which was 48,000,000 pounds. The rapid growth of copper production was due in considerable part to the development in electrical construction, for which copper is peculiarly suited. It is also much used as a constituent of the alloys, —bronze and brass, — and as sheet copper for many purposes.

The United States produces more than half of the world's copper and exports largely. Arizona, Montana, and Michigan are the leading regions. Michigan began to produce in 1845 and has contributed more than any other state, though its annual output is now somewhat below that of Arizona and Montana. Among the companies mining copper in Michigan, the Calumet and Hecla produces the most. The copper in Michigan is "native," that is, not in chemical union with other substances. In this respect the Lake Superior region is said to be "unique among the copper regions of the world." Some workings in Michigan have now attained the great depth of one mile. Copper mining began in Arizona in 1873, while Montana began to be an important producer in 1880. Butte is the center of the copper district of Montana. Utah, Nevada, and California are also important producers, and considerable is mined in Idaho, New Mexico, and Alaska. Several Appalachian states have copper mines, those of Ducktown, Tennessee, alone deserving mention here.

144. Lead, zinc, tin, and mercury. The world's lead product is very small compared with that of copper, and amounted in 1907 to a little less than 1,000,000 pounds, of which the United States furnished about one third. Spain is the chief foreign producer. Both here and in other lands the reserve of lead seems to be small. Missouri, Idaho, Utah, and Colorado are the chief states for this metal. Joplin is the center for Missouri and Cœur d'Alene for Idaho. The use of a large amount of lead for paint is peculiar to the United States, a significant fact, considering that, in the face of growing demand, production is not likely to increase, and also that no lead so used can be recovered, as is the case with much scrap iron and copper.

Missouri mines half of the zinc produced in the United States, followed by New Jersey, Colorado, Wisconsin, and Kansas. As with lead, Joplin is the center in Missouri, and the mines of New Jersey are at Franklin Furnace. Small amounts of tin occur in the Black Hills, and in North and South Carolina and Alaska, but most of the domestic supply is imported from the Malay Peninsula. The making of tin plate from foreign metal is a considerable industry.

Mercury is one of the minor metals, yet its uses are important. It is employed in making thermometers and barometers, in silvering mirrors, in extraction of the precious metals, and in medicine. The only important locality in the United States is in California, though Texas has become a small producer. Spain is the chief foreign source of this metal, and is the only country which mines more of it than our own.

145. Analysis of chief mineral substances in reference to their uses. Water is here included, but will be treated in the following chapter.

Fuel	{ Coal and peat / Petroleum / Natural gas
Nutrition	{ Water / Salt / Fertilizers (indirectly)
Building	{ Stone / Brick / Cement / Glass / The metals, — iron, copper, lead, tin, zinc
Roads	{ Gravels / Paving stones and brick / Asphalt / Cements
Utensils and objects of art	{ Nearly all common metals / Gems and other ornamental stones / Pottery clays / Glass
Record and communication	{ Iron, copper, silver, gold / Bronze, brass / Graphite / Clays / Lithographic stone / Stone and bricks used for inscriptions

146. Precedence of the United States in mineral industries. The total value of the mineral products of the United States for 1907 was $2,069,000,000. Iron, copper, and coal are the three minerals which seem essential to industry, and in these three this country far exceeds any other. Lead, silver, aluminum, and petroleum are also more largely produced here than elsewhere, and the production of gold nearly equals that of any other land.

147. National and state surveys. The United States Geological Survey applies a large part of its energies to the study of mineral deposits and to the dissemination of information. Its reports take the form of bulletins, which give prompt notice of new discoveries; annual summaries of progress in mineral industries; extended monographs and final reports on important regions; and an extended series of papers on water supply. Many state surveys and bureaus coöperate in this work. While mineral products fill the chief place in these reports, much may be found concerning agriculture, plant and animal life, roads, climate, and the forms of the land.

CHAPTER XI

148. Sources of water. All the fresh water that is available for use has come down in the form of rain or snow. The rainfall map of the United States (fig. 102) shows where precipitation is light and where it is heavy. The greatest rainfall is found in the Pacific coast mountains and among the southern Appalachians. The least is found among the plateaus of the western half of the country. All other regions are intermediate.

The water may be directly gathered, for domestic use, in cisterns. In great part, however, it is found in rivers, lakes, underground waters, and springs. Water is always held in plants, animal tissues, and in the soil and rocks. On penetrating below the surface of the land, at a depth of a few or many feet, a point is reached at which all interspaces of soil or rock are full of water. This is a point in an undulating water surface which extends everywhere beneath the land surface. This water surface is known as the water table. Its position depends on the shape of land surface, texture of rock, and changes of season, but always it is the top of an underground reservoir which supplies rivers, springs, and wells.

149. Water as a natural resource. In early days of the white man's life in America, water, like the atmosphere, the soil, and the forest, was thought inexhaustible. Springs, brooks, and wells supplied enough for household use, and any small stream with a swift flow could turn the wheels of a gristmill or saw a few logs for home consumption. There were no large towns whose wastes would pollute the waters, and indeed the dangers of impure water were not known. It is easy to find persons at the present time who are ignorant of these vital matters, and who think one water as good as another if it be clear, cold, and free from bad taste or odor.

The new view of water as a great natural resource, more important than gold, or coal, or iron, has arisen in a time of growing density of population, when large cities require much water for the household, for fire protection, and for manufacture, and when the disposal of sewage and manufacturing waste is likely to contaminate the ground waters and the streams. The dangers

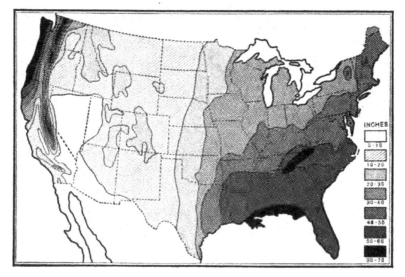

FIG. 102. Mean annual rainfall in the United States

to health and life, due to use of bad water, are also well understood by the greater number of people. Other uses of water also press into the field, and it is not possible to question the statement with which a government expert begins a recent water-supply report: " The water supply of the United States is of more importance to the life and pursuits of the people than any other natural resource."

150. Uses and conservation of water. In the early days of small domestic use water served chiefly for navigation by the small craft of the explorer and pioneer. Irrigation was not needed while population was small and well-watered lands were plentiful. Now irrigation is an absolute essential to the growth of several great states, and is no longer considered unimportant in states having an average rainfall. Probably not one thousandth of the power capacity of our streams was used in the days of small local mills and factories. In the present time, when enterprises are vast, when the fuel should be conserved, and power can be sent far by electric transmission, the use of water power will be pressed to its limit. The ice product, both

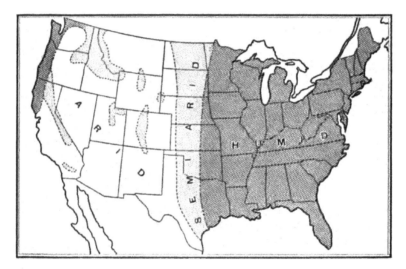

FIG. 103. Humid, semiarid, and arid regions of the United States

natural and artificial, requires a large and increasing use of water. Flood prevention is a part of the proper handling of water resources, and this has to do, in turn, with forest preservation, soil conservation, the promotion of water power, and uniform flow for navigation. Thus nearly all the conservation problems which now engage the public interest are related to the management of water.

Most important of all is domestic and municipal use, — for the household, for manufactures, fire protection, street flushing, public baths, and in relation to parks and playgrounds. The North American Conservation Conference recently laid down the principle that "the highest and most necessary use of water is for domestic and municipal purposes. We therefore favor the recognition of this principle in legislation, and, where necessary, the subordination of other uses of water thereto."

151. Ordinary wells and springs. The essential fact here is that all such sources should be used with caution. Springs and wells that once offered healthful waters may now be full of danger because of growing population and the entrance of household or barnyard wastes into the ground waters that furnish the supply. A student of water supply noted, in an eastern state, between forty and fifty wells located on slopes below cesspools. Under such circumstances even deep wells would be unsafe. Wells 200 feet deep in a New England city showed contamination. It is thus clear that the supply for large bodies of people must be sought at some distance. Great systems of transportation and distribution must be devised, and these constitute large geographical facts of daily importance to the people. Thus the whole question of water has its rightful place in commercial geography.

152. Artesian wells. The artesian well is variously defined. Formerly the term was quite generally restricted to flowing wells. It is now, however, often used of any deep well, even though the water must be pumped from far below the surface. It is better to apply the word to any well in which the water is under hydrostatic pressure, whether it rises to the surface of the ground or not.

Conditions favorable to such wells are found in nearly horizontal and unbroken rock strata, in which some of the series are porous, as, for example, sandstone; while others are compact, serving to hold the water in the porous beds until it is released by well borings. For a fuller account of the nature of artesian wells the student is referred to physical geography.

The Atlantic and Gulf coastal plains and the Great Plains leading up to the Rocky Mountains offer the most widespread favorable conditions for artesian supplies. Thus in South Dakota, according to Darton, such beds as are described above descend

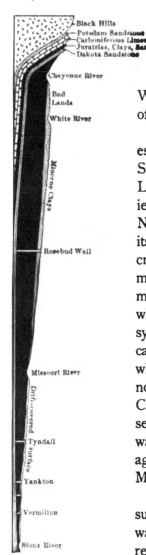

FIG. 104. A diagram illustrating the principle of the artesian well, as shown in South Dakota, from the Black Hills 300 miles eastward. Vertical scale exaggerated

eastward from sources in the Black Hills and carry waters under great "head" to eastern South Dakota and Nebraska. Water when struck often shows a pressure of 500 pounds per square inch.

153. Lakes. Lakes are among the greatest sources of water in the northern United States. Chief among all are the Great Lakes, furnishing navigation, water for cities, and power, as at St. Marys Falls and Niagara. Cleveland, on Lake Erie, draws its supply from the lake by means of two cribs, — one, the older, one and one-half miles from shore; and a newer crib, four miles out. Tunnels transmit the water, which is pumped into reservoirs, the whole system being the property of the city. Chicago has a famous water system of this sort, which has in recent years come into wide notice by reason of the Chicago Drainage Canal. As there was danger that the city's sewage would pollute the waters, the canal was primarily constructed so that the sewage might be turned into the Illinois and Mississippi rivers.

Lake waters are most favorable for a city supply when the shores can be suitably watched and are without large towns. In recent years the city of Syracuse discarded a more local supply and now draws its water from Skaneateles Lake, about twenty miles distant. Innumerable small glacial lakes and ponds, numbering many thousands in the Northern states, now serve or may serve for the supply of adjacent towns and cities.

154. Rivers. Rivers, including all streams large and small, supply a great number of towns with water. Large cities are apt to grow on the lower or middle parts of a river, and there is always danger of water pollution from towns farther upstream.

The subject of river pollution has now entered into law, and in some states, as Massachusetts, Connecticut, New York, New Jersey, Pennsylvania, and Minnesota, the statute restrictions are severe. There is a steady growth of public opinion, enforced by a pronounced awakening on the subject of public health. It is a well-settled part of the common law that a riparian owner has the right to have the water of the stream pass on to him in practically full volume and in normal purity. There is no inherent right to dump sewage or other waste into a river. If pollution makes a stream a public nuisance, all people in the neighborhood as well as those owning the banks of the stream have a right to redress.

In some arid states, where irrigation has developed, the right of "prior appropriation" is established, and late comers find their rights modified by those of earlier settlers.

Government experts have given much attention to the pollution of rivers, and we have examples of numerous cities which have had disastrous experience and have taken effective measures for protection. Among these are Lowell and Lawrence on the Merrimac, Newark and Jersey City, Albany, Pittsburg, Cincinnati, and Louisville. A Hudson River town put a sand filter in operation in 1899, and its death rate from typhoid fever at once dropped to less than one third of that previously registered. This serves to show that the waters of the upper Hudson and Mohawk had become seriously polluted by sewage and other waste, and now nearly if not quite all of the towns on these rivers have adopted upland supplies of water.

A river in the Atlantic coast region became, a few years ago, so foul that two cities were compelled to seek other waters for domestic supply, fish in the stream nearly disappeared, the river could not be used for pleasure boating or bathing, and on certain occasions factories ceased to operate on account of the

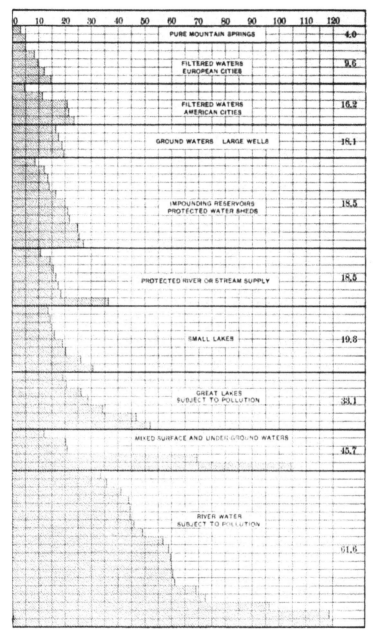

FIG. 105. Mean death rates from typhoid fever in 7 foreign and 66 American cities grouped according to the sources and quality of their drinking water. The first seven are foreign

186

noxious odors. This is an extreme case, but will be only one of many as population grows, unless suitable prevention be adopted both by sewage disposal and by seeking and protecting other sources of supply.

The Connecticut offers an instructive case of a stream useful in several ways. The upper river furnishes wholesome supplies for household use. When it reaches Massachusetts and Connecticut its waters suffer in purity from the large populations on its banks. This is not a serious matter, however, since ample uplands are at hand to furnish waters for domestic use, and the river here offers resources of power of the first order, and in its lower or tidal section is much used for transportation.

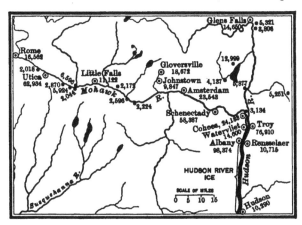

FIG. 106. The Mohawk and Middle Hudson rivers, showing the cities and larger towns and their population, in relation to the ice fields of the Hudson. Cities are named; population only is given for towns

Perhaps the most thorough discussion ever given to river pollution by sewage was during a suit brought before the Supreme Court of the United States in 1900 by the state of Missouri against the state of Illinois and the Sanitary District of Chicago. It was claimed that the discharge of Chicago sewage into the Mississippi River, by way of the Drainage Canal and the Illinois River, was injurious to the people of Missouri, and particularly to those of St. Louis, which draws its domestic waters from the river. The case resulted in a decision rendered in 1906, dismissing the complaint. Among other things it was shown that dilution of the sewage by a large volume of pure water turned in from Lake Michigan improved the water of the Illinois River and caused

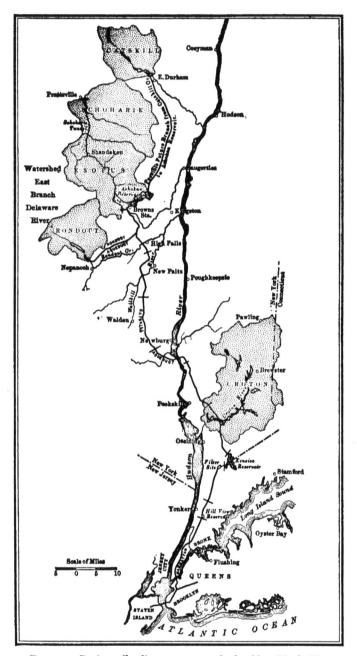

FIG. 107. Basins affording water supply for New York City

188

edible fish to return to it. This decision does not prove that sewage does not contaminate, especially within shorter distances, but only that injury was not proved in that particular case.

The Manhattan and Bronx boroughs of New York are supplied from the Croton and Bronx rivers, with storage reservoirs at High Bridge, in Central Park, and elsewhere. The Croton River is 60 miles long, drains an area of 340 square miles among the uplands of southern New York, and now has 15 storage lakes and ponds. As the city was outgrowing this supply and it was not easy to take water from Connecticut or New Jersey, and as the Adirondacks were distant, an area in the Catskills was chosen, west of Kingston, on Esopus Creek, over 80 miles from City Hall. Here is the great dam for the Ashokan Reservoir, and thence a large aqueduct, built both over ground and with deep siphons, will conduct the water to the city. One great siphon conducts the water beneath the Hudson River at Storm King. The Brooklyn supply is from streams, ponds, and wells on Long Island.

New York and Chicago may be taken as types of great cities, one drawing its supply from small streams and great reservoirs in the uplands, and the other deriving its supplies from a great lake and guarding its purity by costly constructions. Until recently the Schuylkill River furnished 90 per cent of the water for Philadelphia, and the Delaware River the remainder. Hereafter the Delaware is to supply 80 per cent, its waters being cleaner, more free from bacteria, and less hard. Several suburban towns, as Camden, on the New Jersey side of the river, sink artesian wells in the strata of the coastal plain.

Detroit has its supply from Detroit River, which furnishes water that has been directly derived from the upper Great Lakes. It is brought from the center of the channel by a tunnel and raised by pumping. Detailed study would show that thousands of smaller cities and towns have, within recent years, taken measures to secure an ample supply of healthful water, and to discourage, and in some cases to forbid, the use of private wells.

155. Ice. The water of lakes and rivers has become important as a source of ice, made largely by natural, but to some degree by artificial, processes. It has been established that bacteria which cause disease may survive long enough in ice to be injurious. Ice is said to be refused by consumers if cut in the main channel of the Connecticut River, although it may be better than the ice cut so extensively on the middle Hudson, or better than that derived from the lower courses of some rivers of Maine.

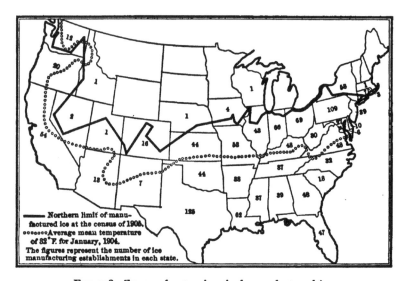

FIG. 108. Zones of natural and of manufactured ice

156. Navigation. This subject is introduced here for a complete general view of the uses of water, but the place of waterways in commercial geography will be discussed in the chapter on Transportation (Chapter XIV). Navigation and power are objects often consistent with each other, but navigation might be unfavorable to domestic uses, as involving more or less contamination. Navigation might or might not be consistent with irrigation, the case depending on the amount of water available and the amount to be taken out for agriculture.

157. Water power. This use of water is important because the supply of fuel is limited. The supply of anthracite coal will apparently be exhausted within the present century, and the bituminous coal in a few centuries. The supply of water power, if wisely used, may be permanent. Water power is more economical also, because it exists in many places to which coal must be brought over long distances. These things give force to the

FIG. 109. Hydraulic laboratory of Cornell University

present movement to keep the water powers from the control of monopolies and to put them under the direction, and make them serve the needs, of the people.

New York may first be taken as an illustration. In an address by Governor Charles E. Hughes before the Conference of Governors held at the White House in 1908, the annual value of the unused water powers of the state was put at $6,600,000. This means that water power not now used in industry would, if employed, earn this sum each year. Added to this present waste is a damage of $1,000,000 per year by floods, which might, by good management, be prevented. The storage of flood waters can best be undertaken by the state, which alone can condemn needed lands and undertake comprehensive operations. The above facts do not mean that New York water power is little used, for in 1905, of all water power employed for manufacture in the United States, New York used one fourth. A water-supply commission has been formed in the state to inquire into the available power, to select good reservoir sites, and to prevent private absorption of these resources. A bill was vetoed, which gave a private company rights without compensation.

FIG. 110. Falls of the Passaic River, Paterson, New Jersey

In a smaller way water powers were much used in manufacture down to 1870. Since that time steam generated by coal has been the chief power, but the new mode of electrical transmission of the energy of both coal and water introduces a new era. Plants now develop power not only for one mill or factory, but for sale to many consumers. The full use of Niagara would give 7,000,000 horse power, and even the present development provides power for many large concerns at the Falls, and transmits tens of thousands of horse power twenty miles to Buffalo and to other points at equal and much greater distances. An expert engineer, also speaking at the White House Conference, places the power of other regions as follows :

Mississippi River system . . .	2,000,000 horse power
Southern Appalachian	3,000,000 horse power
State of Washington	3,000,000 horse power
Northern California	5,000,000 horse power
· United States (entire)	30,000,000 horse power

Storage of waters would much increase these amounts. In 1908, of 30,000,000 horse power used as "prime movers," 26,000,000 were produced by steam engines and only 3,000,000 by water. This shows the vast extent of power now going to waste.

Great development of power is also possible as incident to the improvement of navigation. The locks and dams built and controlled by the government in aid of navigation already afford some 1,600,000 horse power. And at the headwaters of an irrigation stream a power can often be developed without loss to agriculture. The larger irrigation canals may also in some cases, it is stated, serve for navigation.

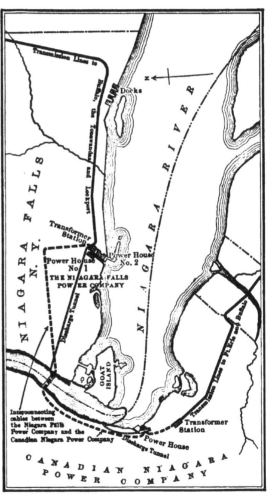

FIG. 111. Map showing transmission lines of the Niagara Falls Power Company .

According to Mr. F. H. Newell, head of the United States Reclamation Service, writing in 1909, the Reclamation Act had been in operation seven years. This act embodied the

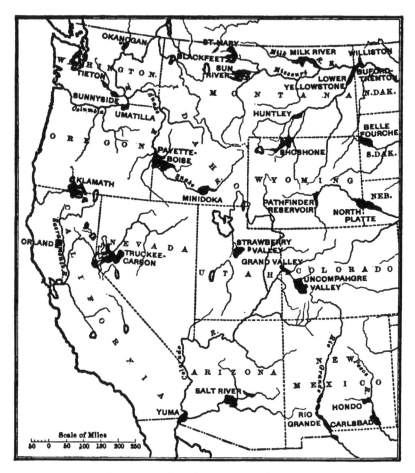

FIG. 114. Location of chief government irrigation projects

first federal activity in irrigation. Works had been built in 13 states and two territories, and nearly 700,000 acres had been brought "under the ditch." Over $1,000,000 had already been returned by settlers, as it is the policy of the government that the entire cost of the works should in time be borne by those benefited. More than 3000 miles of main canals and minor ditches had been constructed.

By consulting the map (fig. 114) the student may see the location of government projects. A few of these will now be briefly described as examples.

Shoshone project, Wyoming. A deep narrow cañon was chosen and the highest masonry dam in the world constructed, 310 feet high. It is higher than the Flatiron Building in New York City, and will create the largest lake in Wyoming; 100,000 acres of land will thus be watered.

North Platte project. In this the Pathfinder dam will create a reservoir for reclaiming 400,000 acres, lying on the trail of the Mormons and the California gold-seekers. This tract is open to settlers in farms of 80 acres. The lake will have more than 10 times the storage capacity of the new Croton Reservoir of New York City. The lands lie both in Wyoming and Nebraska, and will be supplied by a canal 95 miles long.

Belle Fourche project, South Dakota. Here will be formed the largest lake in the state, to irrigate 100,000 acres. Settlement is well advanced, and markets will be found among the mining towns of the Black Hills, in Minneapolis, St. Paul, Omaha, and Chicago.

Rio Grande project. The river of this name is dammed near Eagle, New Mexico, forming a reservoir 175 feet deep, 40 miles long, and sufficient to irrigate 180,000 acres.

Yuma project, delta of Colorado River in Arizona and California. About 100,000 acres of rich soil will be reclaimed, and it is expected that a plot of 10 acres will suffice for the support of a family.

Salt River project, Arizona. Here the government has built the Roosevelt dam, 284 feet high, 1080 feet long, and 170 feet thick at the base. Enormous works will distribute the water, and the climate assures the growth of grapes, almonds, dates, and other products appropriate to warm climates.

Uncompahgre project. The Gunnison and Uncompahgre rivers are 10 miles apart, separated by a mountain range 2000

feet high. A tunnel has been opened 6 miles long, with a cross
section 10½ by 12 feet. The waters of the Gunnison will make
productive 140,000 acres in the Uncompahgre valley.

Yakima valley project, Washington. Several dams provide
water for 500,000 acres. Here very valuable orchards are

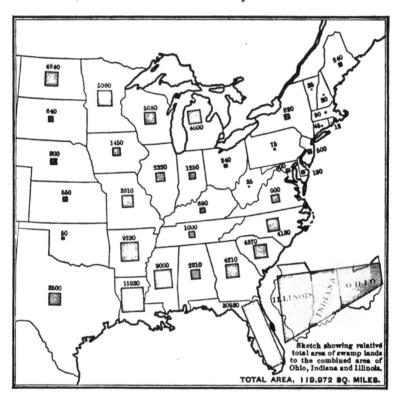

FIG. 115. Areas of swamp and overflowed lands in the United States east of
the Rocky Mountains, compared with the area of three states

developed, the culture is intensive, and farms of 10 to 20 acres
are common. As in many irrigation communities, social contact
is close and the advantages of towns are realized by the people. A further example of this swift development is here given,
selected from many noteworthy cases in the western region.
In southern Idaho a tract on the Snake River was a complete
desert in 1905. In 1908 there were 300 miles of canals and

ditches, 85,000 acres were watered, towns had grown up, and there were 1400 families on farms.

Thus through private and government agency millions of people will be added to the population of the West, the national wealth will in a few years be greatly increased, and the geographic conditions will be transformed through this use of water resources.

159. Prevention of floods and drainage of swamp lands. These two objects form a part of the comprehensive treatment of the waters of the United States. Floods have in recent years

FIG. 116. Traction ditching machine, experimental farm, Cornell University

caused many millions of dollars of damage in the southern Appalachians. They are so severe and so destructive to the cities and farms of the Ohio and lower Mississippi rivers that the United States Weather Bureau maintains a system of flood stations and issues warnings as occasion requires. Other regions suffer in like manner, and it has already been seen that the storage of waters for irrigation and navigation, as well as for the protection of forests, tends to avert floods.

Swamp lands of tidal origin abound on the Atlantic coast. Other swamps are found along the rivers of the coastal and

Gulf plains, and numerous marshes in the Northern states are due to glacial obstruction of drainage and to the filling and partial drainage of lake basins. As population increases, these lands will be reclaimed for suitable additions to the food re- sources and home sites of the people. It is believed that 70,000,000 acres could be cultivated, adding $700,000,000 to

FIG. 117. Gauging river flow, West Carson River, California

the national resources. Extensive drainage operations, at vast cost, are now in progress in the state of Iowa.

160. Mineral and thermal waters. Waters in their sojourn on and in the earth take up mineral substances. There is no exception to this rule, but only when the quantity is pronounced, and especially when the waters are deemed medicinal, is the term "mineral water" applied in a commercial sense. About 10,000 mineral springs are known and recorded in the United States. Most of the springs which supply waters to trade are found in the Appalachian region and in the Mississippi Valley. Most of the remainder are in the Pacific coast states. This does not mean that such waters may not be just as abundant in the intervening plateaus and mountains, remaining to be developed as population becomes larger in that region. The springs of Saratoga, New York, hold first place in number and importance. Thermal springs are found in Virginia and Arkansas, but are more abundant in the West, a fact to be attributed to more recent volcanic activity in that part of the country. In the year 1904 but 484 springs reported actual sales of water, with a total value of product of about $10,000,000.

161. Governmental investigation and use of water resources. For many years the United States Geological Survey has conducted studies of American waters, gauging the flow of all important streams and publishing numerous reports. Records of stream flow have been made at 14,000 stations in the United States. In later years these reports form a series of water-supply papers, numbering into the hundreds. At the same time the Weather Bureau records at numerous stations the amount of rainfall. The Reclamation Service utilizes water for agriculture in the manner already described, and the more recently created Waterways Commission is charged with development of plans for a suitable system of interior navigation.

CHAPTER XII

162. Beginnings of industry. Before there were civilized peoples there was little diversity or specialization in industry. Each family or tribe supplied its own needs. There could be no geography of commerce because there was no commerce. When tribes began to barter, they had found that others had or made what they did not possess or produce, and thus there was a rude concentration of resources and handicrafts in certain places. As the world of groups that knew and dealt with each other grew larger, men learned where the best things were made and where their products could be sold, and commerce became fully established. The very existence of commerce, therefore, depends on concentration, or at least a grouping, of industries.

When we come to the modern world these processes of specialization and adaptation to particular surroundings have gone far. In Europe, England very largely supplies the textiles, Germany the chemicals, France and Italy the wine and silk, and Russia and Roumania the wheat. If the United States be added to the list, the group of exchanges and adjustments is still wider, and cotton may be taken as an example of a product little raised in Europe. As China and the whole Orient open, and as Africa is subdued and civilized, and South America comes fully into the family of industrial nations, the world becomes a patchwork of products and industries and a network of lines of exchange.

The same principle applies to a single country. In England, Lancashire makes cotton goods, Sheffield forges iron and steel, Leeds furnishes woolens, and Newcastle ships, while Norfolk raises wheat and Kent produces hops. This chapter will review the special location of industries in the United States, observing the simplest and closest concentration of particular manufactures. Not only the place, but the reasons for the place, will be sought.

163. Cases of concentration already noticed. Minneapolis doubtless became the place for wheat milling on a large scale in the Northwest, because of the power supplied by the Falls of St. Anthony. This cause is not now so important as in the beginning of the industry. Minneapolis is also easily reached from the present wheat center of the United States, and the most perfect modern transportation has been developed in all directions. Further, Minneapolis lies en route to the wheat and flour markets of the East and of Europe. And yet again, this city has established a reputation for the finest brands of flour, its trade-marks are accepted the world over, much capital has been invested, and skilled labor has developed. If some other place could be found having greater natural advantages, the business would persist here because it is so firmly fixed in its appointments and in the minds of the trade and of the people at large.

There is a striking concentration of cotton manufacture in coastal New England, especially in the regions of southwestern Maine, southern New Hampshire, eastern Massachusetts, Rhode Island, and eastern Connecticut. This was favored in early days by water power, by a moist climate, and by available labor, intelligent and faithful operatives being drawn from New England towns and farms. At the present time water power is of much less account, though Lowell and Manchester still use it for about 50 per cent of their cotton manufacture. Artificial moisture is now preferred, and people from other lands, particularly the French from Canada, contribute largely to the labor supply.

There is in New England, around the cotton industry, an intrenchment of capital, reputation, and skilled labor which tend to hold the industry. The region is also near the largest markets. On the other hand, the raw cotton and the fuel must be hauled for long distances. Cheaper labor, water power, and the nearness of the cotton favor the migration of the business to the South. Massachusetts therefore, while holding 37.4 per cent of the American cotton manufacture in 1890, held but 32.8 per cent in 1900. In like manner Rhode Island dropped from 10.2 per cent to 7.8 per cent in the same decade. There was a relative decline in every New England state, while South Carolina was the second state in 1900, having 8.8 per cent as compared with 3.7 per cent in 1890.

The first five cities in 1900 were Fall River, Philadelphia, Lowell, New Bedford, and Manchester. Warwick, Rhode

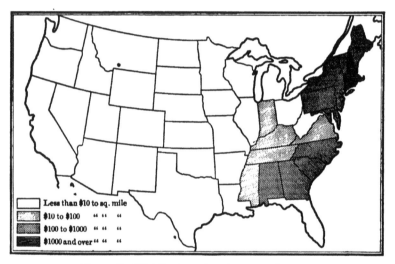

FIG. 118. Distribution of cotton manufacture

Island, was the most highly specialized town, cotton forming 71 per cent of its manufacture. The percentage for Fall River was 68, for New Bedford 65, and for Lewiston, Maine, 54. The cotton industry represented but 2 per cent of the manufactures of Philadelphia, the second producer, because this is a great general center of industries, while the others are small cities largely given to one pursuit.

Cattle and hogs are chiefly raised in the Mississippi basin. The greater markets for dressed and packed meats are in the East and in Europe. The centers for slaughtering and packing would naturally be near the farms and ranches which produce the stock, or on the way toward the markets. As a matter of fact they are nearer the farm than the market, at least in Omaha, St. Joseph, and Kansas City. The meat can be better shipped than the stock " on the hoof," although much is carried in the latter way. Further, such a by-product as fertilizer may better be made where reshipment to the farm is easy and cheap. The centralizing of meat packing is only possible through preservative processes and swift and effective transportation.

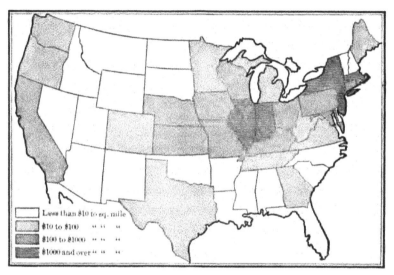

FIG. 119. Distribution of slaughtering and meat packing

The two conspicuous concentrations of iron making are found in the southern Appalachian valley from Knoxville to Birmingham, and in the Pittsburg region from western Pennsylvania extending into Ohio. The first illustrates the effect of having all the raw materials close at hand. The second has the fuel at the door, and the iron ore, while far away, is made most accessible by the best modern transportation. The same is true of Buffalo and Chicago, though in these cases fuel also is carried some distance. The Northern centers named are nearer than the Southern centers to the great markets, for the various steel products, which are found in the densely populated Northern states.

Fifty-six per cent of the iron and steel made in the United States in 1900 was made in Pennsylvania, and 11 per cent was produced in Pittsburg. Pennsylvania has always been first in iron, but the use of Connellsville coke and Lake Superior ores shifted the center from eastern to western Pennsylvania. Eastern Ohio really belongs to the same region, and state boundaries should be little heeded in this study. Thus, in 1900, the region actually produced 71 per cent of our domestic iron and steel. The following figures show the proportion of iron to other manufactures in certain towns:

McKeesport, Pa. . . . . 92 per cent
Youngstown, Ohio . . . . 81 per cent
Johnstown, Pa. . . . . 79 per cent
Joliet, Ill. . . . . 48 per cent
Pittsburg, Pa. . . . . 44 per cent

The low percentage for Pittsburg is explained by the fact that a great city must always have a considerable variety of industries, even though best known by one.

164. Causes of concentration. The more common causes for "localization of industry" are summarized by a writer in the census of 1900 as follows:

1. Nearness to materials (including fuel)
2. Nearness to markets
3. Water power
4. Favorable climate
5. Supply of labor
6. Capital available for investment
7. Momentum of an early start

These principles having been stated, other industries will be taken up and the principles applied. It is useful to the student of commercial geography to know first the facts of distribution, that is, where industries are; and second, to inquire why they are so placed, with care to see that commonly not one cause but two or more are or have been influential in varying degrees.

165. Knit goods. In this industry there is considerable concentration. A visitor to the towns of the Mohawk and middle

Hudson rivers in New York would find many large mills and a great body of operatives trained to the use of highly elaborate machinery. The Empire State makes more than one third of the knit goods manufactured in the United States, and is the center, although Philadelphia is a larger producer than any other single city. Cohoes and Amsterdam in New York are second and third, and Lowell, Massachusetts, is fourth. Cohoes and Amsterdam differ from both Philadelphia and Lowell in being

FIG. 120. Operatives in knitting mill, Amsterdam, New York

more especially devoted to the knitting industry, the share of the total output of local industries being 43 per cent in Cohoes, 37 per cent in Amsterdam, and but 2.6 per cent in Philadelphia. As before, it is a difference between a center of special industry and a center of general industry. This type of manufacture began in Philadelphia in 1698, with the coming of expert knitters from the German Palatinate. The industry in eastern New York is much younger. In 1832 a Cohoes manufacturer invented a power knitting machine, which largely promoted the industry. In Cohoes 75 per cent of the power used is derived from the Mohawk River, whose waters fall 70 feet at that place. Amsterdam, on the Mohawk, 33 miles from Albany, appears to have developed the industry because of nearness to Cohoes, having a small water power but chiefly using coal. This industry is now extensive in the cities of Utica and Little Falls in central New York.

166. Collars and cuffs. This product offers the most complete case of concentration supplied by the industries of the United States. The whole product in 1900 was valued at $15,769,000, of which the state of New York was credited with $15,703,000, or 99.6 per cent. Of the total, the single city of Troy produced 85.3 per cent, most of the remainder being made in Glens Falls. Collars and cuffs provide almost half of the industry of Troy. The last cause of concentration stated

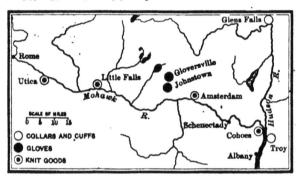

FIG. 121. Centers of production of collars and cuffs, gloves, and knit goods in eastern New York

in the census report as above quoted, applies forcibly here. There is no reason in the location itself why collars and cuffs should be made in a particular town on the Hudson River, rather than at any other place accessible to eastern markets. There was an early and favorable start at Troy, and the business there gained the reputation of being profitable. Detached collars and cuffs are said to have been invented there, and this added popularity to the trade. Skilled operatives were trained, machinery was invented, and buyers formed the habit of looking thither for their supplies. Capital interested in this industry went there, rather than experiment in a place unknown to the trade. The very atmosphere and life of the town became pervaded by this single industry. Laundries in many towns were called Troy laundries, to take advantage of the fame of the place in this respect.

167. Gloves. In this product New York supplies another conspicuous illustration of concentration. Gloversville and Johnstown are small cities in Fulton County, New York, and are

FIG. 122. Preparation of skins, Gloversville, New York

situated a few miles north of the Mohawk River, close to the southern border of the Adirondack Mountains. The following statement is taken from the census report of 1900 :

> 381 glove factories in the United States
> 243 glove factories in the state of New York
> 166 glove factories in Fulton County
> 101 glove factories in Gloversville
> 49 glove factories in Johnstown
>
> $16,000,000, total product in the United States
> 9,380,000, total product in Fulton County
> 6,350,000, total product in Gloversville

Thus a single small. city of 20,000 people made about 40 per cent of all the gloves produced in the United States. Johnstown adjoins Gloversville, although a separate municipality. Both may properly be taken as one center, and thus the percentage of the total becomes nearly 59. The whole industrial life of the two towns is centered in gloves. In the lobby of the chief hotel one may see glove and leather men from all states and all lands. On almost every street are factories, some large, and often small establishments in the rear of dwellings. Racks are hung with skins in various stages of preparation. Everywhere in the surrounding country for a dozen miles sewing machines are in operation in country houses, where women add to the income of the household by sewing the gloves cut in town and distributed and

FIG. 123. Cutting room of glove factory, Gloversville, New York

gathered by teams at intervals of a few days. A few factories are located in adjoining villages, and the whole region is given to the business.

Localization is not due to any nearness of raw material or of freight, for these towns are subject to the rates of a single local railway company. Freights, however, are less important for gloves than for products of greater weight or bulk and less value, such as grain, coal, or crude ore. Again census cause number 7 is all-important, — the momentum of an early start. Sir William Johnson brought from Scotland emigrants skilled in this art. They had been members of the glove guild in Perthshire, and a small town not far from Johnstown now bears the name of Perth. In 1809 gloves were first taken to a market beyond the home neighborhood. A bag of gloves was taken to Albany and sold at a profit, and so the industry was stimulated. In 1825 Elisha Judson took a lumber wagon loaded with gloves to Boston. He was gone six weeks and brought back to his employers $600 in silver coin, whereby the business received another impulse. Foreign skilled workers continued to come in, the native inhabitants learned proficiency, capital sought the place, and the trade centered here. It is said to be difficult to

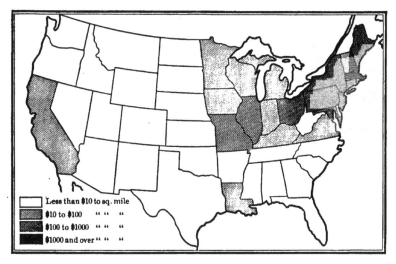

Less than $10 to sq. mile
$10 to $100   " "   "
$100 to $1000   " "   "
$1000 and over " "   "

FIG. 124. Distribution of the manufacture of boots and shoes

persuade skilled glove makers to go elsewhere. Chicago produced in the census year gloves to the value of over $2,000,000, and thus ranked next to Fulton County, but this would be but a small incident in a great industrial focus like Chicago.

168. Boots and shoes. Eastern Massachusetts is the American center of this industry, and had nearly all of it until the nineteenth century was far advanced. In the census of 1900 Massachusetts reported $117,000,000 out of a total of $261,000,-000. There have been recent advances in other states, but this progress in some degree belongs to the adjoining parts of New Hampshire and Maine. In 1890 Massachusetts made 53 per cent of the product, but in 1900 had dropped to 45 per cent, showing a tendency to diffuse the industry. The concentration in cities is marked. Brockton, Massachusetts, was first, Lynn second, and Haverhill third. Men's shoes prevail in Brockton factories and women's shoes in those of Lynn, showing still closer specializing. Many shoes are made in Philadelphia, Brooklyn, Rochester, Cincinnati, Chicago, and St. Louis, but none of these cities made much more than half of the value credited to any one of the three leading shoe towns of Massachusetts.

The industry began in Lynn in 1636. The shoes of the continental army were made in Massachusetts and were considered as good as if made in England. To-day the United States sets the standard for the world in shoes. In 1795 Lynn had 200 master workmen and 600 journeymen and turned out 300,000 pairs of women's shoes. Eastern Massachusetts has had the "momentum of an early start." It was long favored by the local manufacture of leather, but the westward movement of leather tanning and dressing tends to draw the shoe industry with it. On the other hand, the local reputation for shoes, the capital invested, and the skill acquired tend to hold the manufacture where it is. The highest degree of absorption in this single industry is seen at Brockton, where the percentage is 75.

169. Silk. Paterson, New Jersey, is the center of silk manufacture in the United States. This does not mean that Paterson makes nearly all American silk, for the percentage of the total in 1900 was 24.2, or $26,000,000 out of $107,000,000. But no other center approached Paterson, the second being New York, with value between 6,000,000 and 7,000,000. The interesting fact about this example of concentration is that it is far from the source of the raw material, which must all be brought across the sea. Paterson is said to owe its rank to its nearness to New York, the greatest American market for silk. It was favored by the water power of the Passaic River and has attracted skilled labor from Italy and other parts of Europe.

170. Glass. Nearly two thirds of the American glass product is made in Pennsylvania and Indiana, New Jersey and Ohio ranking third and fourth. Good fuel is the essential factor in glassmaking. Coal and gas are used in Pennsylvania, and the development in Indiana has depended chiefly on supplies of natural gas, which is especially adapted to this manufacture. The two city centers are Pittsburg, Pennsylvania, and Muncie, Indiana, but neither shows much more than 4 per cent of the total. Many small towns in these regions have an important share in the industry and contribute the greater part of the product. Glass sands are not uncommon and are sometimes derived from unconsolidated and recent formations, but often from ancient and compact sandstones. The localization is dependent more on fuel than upon the sands. The glass industry began in early colonial days and was established beyond the Appalachians at Pittsburg so early as 1796. In recent years there has been great advance both in quantity and quality of the product, as well as a decrease of importations.

171. Pottery. Ohio, New Jersey, and Pennsylvania, in the order given, are the first producers, making two thirds of the domestic product. Trenton, New Jersey, and East Liverpool, Ohio, are the chief town centers. There are forty potteries in Trenton, where the industry began at an early date. The largest single pottery in the United States is at East Liverpool, which was at one time almost an English town, owing to the coming of so many skilled workers in pottery from the mother country. The location of the industry at these two points does not seem to be due to the presence of raw material, except as coarser clays favored the very beginnings. Then skilled labor was attracted, and clays for the finer wares were brought from a distance. English clays, for example, are brought to Trenton at low cost because ships bearing heavy cargoes eastward are glad to ballast at low rates for the westward voyage. It is said that it costs as much to bring a ton of clay by rail from beds in New Jersey and Delaware, or even "sagger" clay from pits four miles away, as it does to transport a ton of the finer clays from England to New York City. Therefore this industry also illustrates the principle that accessibility

by ship or rail may be as favorable as actual nearness of raw
material or market.

172. Agricultural implements. Illinois, Ohio, and New York
hold the larger part of this field of manufacture, but Illinois
produces twice as much as the other two states combined, its
output amounting to $42,000,000, or 44.5 per cent of the
total. There has also in recent years been an increase in Illinois and a decrease in Ohio and New York, thus showing

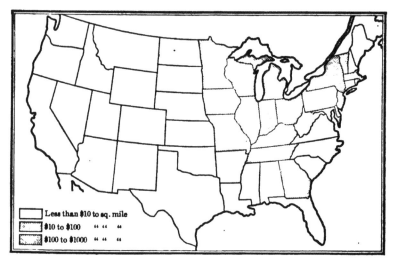

FIG. 125. Distribution of the manufacture of agricultural implements

that the industry is moving westward, following the movement
of agriculture in that direction. Chicago is the greatest city
center, producing to the amount of $25,000,000, or more than
five times as much as its nearest rival, Springfield, Ohio. Chicago is near the grain fields, and is well placed for supplies
of coal, hard wood, and iron. Both raw materials and markets
are either near, or are made accessible by the best means of
transportation. In 1880 Springfield produced as much as Chicago. Being a small center largely given to one industry, 41.3
per cent of its manufactured output is agricultural machinery,
while Chicago shows but 2.8 per cent, because it is a vast
center of general industry. Racine, Wisconsin, Auburn, New

York, and South Bend, Indiana, are highly specialized, but to a less degree than Springfield.

173. Center of manufactures. The foregoing examples sufficiently illustrate the principles involved in concentration. Changes in any of the factors which lead to the concentration of particular industries (sect. 164) will tend to shift the special centers. Thus the development of water power, of skilled labor, and of local markets aids in moving cotton manufacture from New England to the South. The westward movement of boot and shoe making and of the manufacture of agricultural implements has already been interpreted in similar ways. Here, therefore, the principle of constant geographic *adjustment* is introduced and illustrated. It may finally be observed that manufacturing as a whole predominates in the eastern United States, where a considerable share of the raw materials is found, where there has been time to develop the required skill, and where the larger populations and greater markets are located. In the census of 1900 the center of population was in Indiana, but the center of manufacture was eastward, in central Ohio.

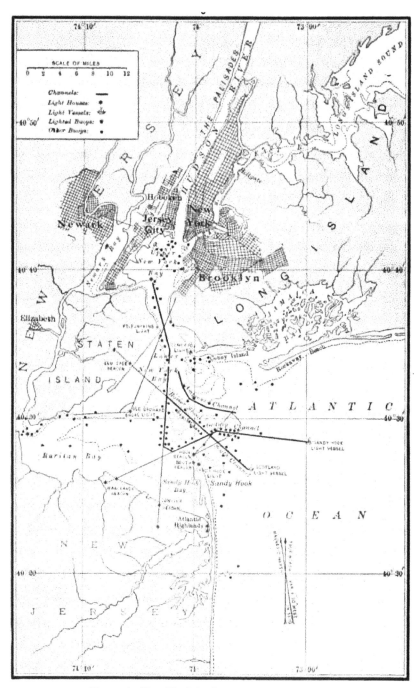

FIG. 126. New York harbor and its approaches

216

CHAPTER XIII

174. Introductory statement. In Chapter XII the concentration of single industries was studied. In some cases the industry made most of the business of a town, as collars and cuffs in Troy or pottery in East Liverpool. In other cases a large concentration of a single interest formed but a fraction of the manufactures of a city, as agricultural machinery in Chicago or knit goods in Philadelphia. Chicago and Philadelphia are therefore general centers, with a large variety of industries and a diverse trade. To the growth of such centers the present chapter is given. New York, Chicago, and St. Louis are taken at the outset, the first being an ocean port, the second a lake port, and the third a river port.

175. New York. This city was first built on an island, a condition useful for defense in early days, and now an advantage for the extent of water front. It is also a hindrance to growth, making transportation to the outlying boroughs and suburbs difficult. This is remedied as much as possible by bridges and tunnels. The harbor is the best in the New World and offers some hundreds of miles of wharfage. The route to Europe is shorter than from Philadelphia and Baltimore, but a little longer than from Boston. The city is as well placed as any rival for the vast coasting trade. The region tributary to a city's trade is often called its hinterland. There is always a local hinterland whose relations to a great city are close, whose gardens and dairies supply much that the city needs, and whose towns and cities contribute to the trade of their greater neighbor.

The highlands come so close to the sea at New York that its local hinterland is smaller than that of Boston, or Philadelphia, or Baltimore. But New York's greater hinterland reaches to the Great Lakes, the prairies, and to the Rocky Mountains. The Hudson River is navigable for 150 miles. The Mohawk valley offers a pass in the Appalachians, at an altitude of but 445 feet. The Erie Canal, finished in 1825, made the rich region of western New York tributary to the city of New York, and the Great Lakes led to the heart of the prairies. Soon the railroads joined lakes, prairies, and Great Plains to New York, and thus, by the North Atlantic, to the markets of Europe. Its place upon the national and international highway is the chief explanation of New York's greatness. It is also favored by the nearness of Pennsylvania coal.

New York passed Philadelphia in export trade in 1789, and at length outstripped its rivals in interior trade because the Mohawk route was easier than those of the Susquehanna and Potomac. In 1823 the foreign immigration to New York exceeded that of its three rivals combined, and in 1850 the city had more than 500,000 people.

Following these causes of growth, it is easy to see how manufactures and commerce grew also. The growing population made an important market at home. Imported raw materials were brought there, manufactured, and sent to interior markets. Interior products came to New York and were shipped to Europe. The more factories there were, the more people were required to operate them. The more goods were transshipped, the larger was the population required for such traffic. The increase of immigration gave the manufacturer an abundant supply of cheap labor, and cheap labor and a large market attracted more capital and more industries. These principles are not peculiar to New York, but operate in favor of every industrial center. Financial, social, educational, and literary advantages attract to every great city, and the opportunity for labor and for amusement brings · many. Thus the local market grows and the relations to surrounding territory become closer.

From 1900 to 1905, there were 1596 new manufacturing establishments started in New York. There were 20,839 establishments in 1905, representing 291 industries out of 339 classifications used by the Census Bureau. New York makes one tenth of all the manufactured products of the United States. Only two other cities had totals which equaled the clothing industry alone in New York. The value of this product for the city in 1905 was $305,000,000. Printing and publishing ranked as the second industry, and no other American city approaches New York in this respect. Twenty-six other industries in New York each reported a product valued at more than $10,000,000, while tobacco, slaughtering, bakery products, and malt liquors each exceeded $40,000,000. Sugar and molasses, copper refining, and petroleum refining aggregated more than $150,000,000, but are not given individually because each was represented by one vast establishment, and it is the policy of the census not to give figures which reveal the extent of the business of a single concern. If all the manufactures of the region closely dependent on a great city are included, the totals are greatly increased, so that the above figures do not fully express New York's industrial greatness. The interests centering in New York make it the financial and banking center of the entire country.

176. Chicago. This center had in its beginnings no local advantages, for its ground was low and wet. It was, however, inevitable that a city would develop at the head of Lake Michigan, and the actual place was determined by the entrance of the small Chicago River, whose lower part affords a certain amount of harborage.

The advantages of the site may be briefly stated. Being at the southern end of Lake Michigan, Chicago is at the head of lake navigation for the central United States. Duluth, at the head of Lake Superior, is too far north, but would, in like manner, command a vast region if the Canadian prairies were not in another country. All eastern land traffic from Wisconsin, Iowa, Minnesota, and the more distant Northwest must pass around Lake Michigan, and hence is tributary to Chicago. In addition, all of the upper Mississippi Valley is accessible to Chicago, and is a region of easy grades for railways, of suitable surfaces for agricultural machinery, and has rich soil and mineral resources.

Perhaps no other city in any land has so important a local hinterland, combined with such facilities for shipment both by land and water. It has been said that Chicago is at the western terminus of the Erie Canal, and it is on the trunk line of communication which leads from western Europe to New York, to the Pacific coast, and to eastern Asia. In addition to these conditions Chicago has been guided and built by men of progress and energy.

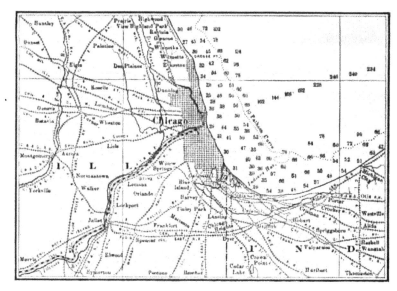

FIG. 127. Chicago and vicinity. Soundings are given in feet

Like New York, Chicago has breadth of industry, but its industries are more dependent on surrounding lands than are those of New York. Chicago had 8159 manufacturing plants in 1905, with total products valued at $955,000,000. Slaughtering and meat packing were first, with nearly $270,000,000, which was not far below the total for clothing in New York. Chicago reported seventeen other industries, with outputs each in excess of $10,000,000. Some of the larger were clothing, foundry and machine-shop products, iron and steel, cars, electrical machinery and agricultural implements, and lumber. Of all thus far named, the meat, metal, and lumber industries depend on the accessibility of raw materials to Chicago. Men's and women's clothing made nearly $65,000,000, a little more than one fifth of New York's product. Printing and publishing amounted to $48,000,000, as compared with about $116,000,000 for greater New York for the same year (1905).

177. St. Louis. This city has the most commanding position of any American center so far as river navigation is concerned. Being on the Mississippi below the entrance of the Missouri and above the mouth of the Ohio, it is peculiarly accessible to the entire Mississippi system of waters and to the lands over which they flow. These lands extend from the Rocky Mountains to the Appalachians, and from the Gulf almost to the Great Lakes. In early days St. Louis was commercially important as a fur depot, and then as a distributing point of supplies for the Southwest, particularly by the Santa Fe Trail. It had a large steamboat trade in the middle of the last century, or until railroads absorbed the traffic. St. Louis was in a good position for a railway center, and it now vies with Chicago as a point of arrival and departure between the East and the West. The first important industry was in meat and flour, and these are still large interests. Southern Missouri is a region of rich and varied mineral resources, which are tributary to the chief town of the state. Close at hand also are the great tobacco fields in the region across the Mississippi to the eastward. There is also timber in Missouri and several adjoining states, and plentiful coal in Illinois and Iowa. The industries are varied, and no single item is as prominent as is clothing in New York or slaughtering in Chicago. As in many other cities, facility of transportation is the largest factor, the city having formerly prospered through river traffic, as it now thrives by railway traffic.

178. Principles affecting the growth of centers of general industry. These principles cannot be fully classified or stated, but the following have special importance:

1. Suitable local site. This is, however, not a controlling condition.

2. Facility of transpórtation. This is the main condition, and belongs to places through which people must pass, and to which or through which they naturally send their products. This central character as regards accessibility of wide surrounding lands belongs to Chicago and St. Louis, and to New York in even a larger way, as being between the United States and Europe. New York harbor, by its excellence, and the Hudson-Mohawk route have determined which Atlantic port should be the greatest.

3. Importance of the hinterland, which is made accessible by good routes. This is involved in the second principle.

4. Conditions for manufacture, such as power, quality of water and of the atmosphere, or the presence of raw materials. Here, however, transportation makes a center less dependent on local conditions.

5. Attraction of cities, with their schools, libraries, amusements, opportunities for capital and labor, and the fascination of stir and of large bodies of people.

6. Political causes. Washington is an example, or Indianapolis. In the latter case the state capital was central to a rich region, has become a great railroad center, and has largely outgrown the fact that it is a state capital. Columbus and Des Moines are similar examples.

All these principles have to do with growth. The more people congregate the more industries must arise ; and, conversely, the more industries develop, the more people assemble around them to man them and be supported by them. If there are facilities for a particular sort of manufacture, then industry becomes specialized, as in Pittsburg ; but never can a great city be so largely given to one industry as is Gloversville to gloves or Fall River to cotton. A great city must have varied industries, a condition which may, or may not, belong to a small city. Some places have been special centers when young, but have wider interests as they grow larger. Thus Rochester was once the " flour " city and Syracuse chiefly made salt, while these are now minor interests, and industry in both places is diversified.

179. Atlantic ports other than New York. Boston, Philadelphia, and Baltimore will be considered. In early days there were several important ports in eastern Massachusetts, as Salem, Boston, and Plymouth. Boston gained precedence as larger ships sought the deepest harbor and the railroads chose it as a terminal. The Boston and Albany Railroad and the Hoosac Tunnel Route gave Boston a share of the Western trade, but it has been truly pointed out that her relation to all New England is the main factor in her growth. The capital and management of most New England mills center in Boston. The leather and rubber industries lead, while slaughtering is third (for the industrial district of Boston), notwithstanding remoteness from the stock-raising region. Fine metallic goods, such as tools, watches, and jewelry, are manufactures appropriate to an old and highly developed industrial community.

 Philadelphia and Baltimore have the advantage of deep tidal waters, bearing the largest ships far inland. At an early date both had well-built earth roads to Pittsburg and the Ohio valley, and these were succeeded by the Pennsylvania and the Baltimore and Ohio railway systems. Both likewise have rich hinterlands in Pennsylvania, New Jersey, and Maryland. Various textile industries and the refining of sugar and petroleum are characteristic industries of Philadelphia, while locomotives and shipbuilding hold no small place. Typical industries of Baltimore are the manufacture of tobacco, and the canning of fruit, vegetables, and oysters.

180. Pacific coast centers. Here good harborage has determined the sites of the chief towns, and has drawn to them the railroads from the East and sent out from them lines of commerce across the Pacific. San Francisco on San Francisco Bay, Portland on the tidal waters of the Willamette and Columbia rivers, and Seattle on Puget Sound are the great and growing centers in this region. All have rich and extensive valleys as a local hinterland, with a larger background reaching eastward to the Rocky Mountains. All are developing Pacific trade, and Seattle especially is brought into close relation with the growing

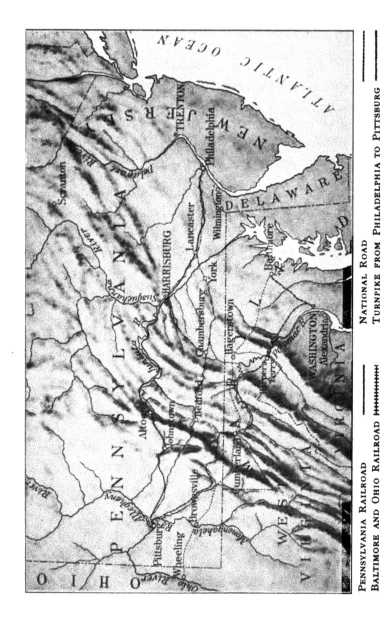

PENNSYLVANIA RAILROAD ————   NATIONAL ROAD ————
BALTIMORE AND OHIO RAILROAD ⊢⊢⊢⊢⊢⊢⊢   TURNPIKE FROM PHILADELPHIA TO PITTSBURG ————

FIG. 128. Philadelphia and Baltimore, with hinterland areas and roads to the west

224

trade of Alaska. There is no great and leading industry in San Francisco, but a variety of manufactures to suit the general needs of the Pacific coast region. Such are the foundry, slaughtering, printing, clothing, canning, and lumber industries. Lumber industries lead in Portland, as might be expected, followed by flour, slaughtering, foundry, bakery, and printing. Similar conditions prevail in Seattle, where, however, flour is the largest item of manufacture.

181. Great Lake centers other than Chicago. Two of these are on Lake Erie, one at its foot and the other on its south shore. Cleveland shows the effect of special transportation relations, which bring to it coal and iron, and the iron industries are foremost, including iron and steel, steel bridges and buildings, metal-working tools and machines, carriage and vehicle hardware, automobiles, vapor stoves, gas ranges, and ships. In addition there are 134 foundries and many machine shops. Cleveland in 1905 was the chief manufacturing city of Ohio, slightly surpassing Cincinnati for the first time. Petroleum refining is another large industry made possible by the neighborhood of a great oil region and favored by pipe-line transportation of the crude oil and water carriage of the refined products.

Buffalo has a rich local hinterland in western New York, and is preëminent in its advantages of transportation, with several trunk lines of railway, the Erie Canal, and the entire sweep of the Great Lakes. If the Barge-Canal enterprise in New York is successful, Buffalo will be yet more favored. The city has also a unique situation for power, by reason of Niagara, which is but 20 miles distant. It is indeed the opinion of certain experts in England that the great manufacturing city of the future will be on the banks of the Niagara River. Flour, slaughtering, and iron are the first items of Buffalo industry, and the last is growing rapidly. It is claimed that more than enough is saved on ore and limestone to offset Pittsburg's advantage in coal, and within a few years iron companies employing thousands of men have established themselves there on the shores of Lake Erie.

Detroit is at a point where the stream of Great Lake traffic is crossed by great railways leading from east to west. Its special hinterland is the southern peninsula of Michigan. It now has a smaller share of lake traffic than formerly, because much shipping passes from Lake Superior to Lake Michigan and to ports on Georgian Bay. The manufactures are widely varied, and Detroit can hardly be said to have a leading industry. Milwaukee has an important tributary region in the great state of Wisconsin, and its leading industry is the making of malt liquors, followed by leather and iron.

182. River centers other than St. Louis. Cincinnati, Louisville, and New Orleans are here taken. Cincinnati is naturally compared with Pittsburg, both being on the Ohio River, though the output of manufactures is much greater in Pittsburg than in Cincinnati. The latter has no such leading industry as ironwork is for Pittsburg. It is also compared with Cleveland, which it about equals in industrial products. Its situation is due to the entrance of the Miami upon the Ohio valley, and, as in every other example, transportation is a main cause for growth. It is on the Ohio River, is the terminus of a canal leading to Lake Erie, has attracted to itself many railroads, and is on the main avenue from the Northern to the Southern states between the Appalachians and the Mississippi River. It has large slaughtering and meat industries, though not relatively so large since the grazing regions have pushed westward. Liquors exceed the animal industries, and Cincinnati, like St. Louis, is on the borders of the greatest American tobacco fields.

Louisville was determined originally by the falls of the Ohio River and the former break in river trade at that point. It was on the line of all early movements westward from Philadelphia and Baltimore, and had river trade with New Orleans and the West Indies. Its industries depend much on the raw materials of the neighborhood. Thus the larger items are tobacco and whisky, the latter being related to surrounding grain fields and to the quality of the water. Cooperage follows upon whisky, and this, with other wood industries, is favored by neighboring forests of hard wood. New Orleans is greater in the commercial than in the industrial way, and ranks at once as a river center and a seaport. The improvement of the Mississippi River, the opening of the Panama Canal, and the new development of the South will all favor its growth and may lead to very great enlargement.

183. Other cities. Omaha, Kansas City, Denver, and Los Angeles are chosen. All of these except Los Angeles are on considerable streams, though Denver is far up the South Platte, and in no case does water transportation play a large part. The first transcontinental railroad crossed the Missouri River at Omaha and made it the chief city of Nebraska. Including South Omaha, it has one great industry, — meat packing, — the product amounting in 1905 to $65,000,000. In like manner Kansas City (Missouri and Kansas) has a large animal industry, but also has many others of moderate size for the supply of a local but important region.

Denver is at the eastern front of the Rocky Mountains, is the capital of Colorado, and is the financial center of the great mining interests of this state. It is also a center of general industry for a wide region of the Rocky Mountains and adjoining plains. Los Angeles, like Detroit, is a center of general industry. As with Denver it is favored by large development of through railway lines, and is the local center for southern California.

184. Industrial districts. The United States census has studied a dozen of the greater American cities by including with the city proper the surrounding region and adjacent towns whose manufactures are so closely related to it as really to belong to it. Viewed in this way, Boston may be regarded as having more than 1,350,000 people, or more than twice the population shown for the city itself. In like manner New York would have at the present time nearly or quite 6,000,000 people. The census of 1910 gives Greater New York a population of 4,766,883. Across the Hudson River in New Jersey is a region of almost continuous settlement, extending many miles from New York. In this suburban region the cities having a population of more than 25,000 each, aggregate about 1,000,000 people, who would naturally be citizens of the metropolis if a state boundary did not intervene. In like manner many towns and cities in New York and Connecticut are close to the greater city.

Twelve such districts may be taken, including New York, Chicago, Philadelphia, Boston, Pittsburg, St. Louis, Baltimore, Cincinnati, Buffalo, Minneapolis, St. Paul, and San Francisco. Considering the 10 leading industries of each for 1905, foundry and printing alone appear in all. Clothing appears in 10, slaughtering in 10, liquors in 9, and tobacco in 5. Slaughtering is first in Chicago, St. Louis, and San Francisco. Clothing is first in New York and Baltimore, and textiles lead in Philadelphia. Iron and steel are first in Pittsburg, Cleveland, and Buffalo, while boots and shoes hold first rank in Boston, liquors in Cincinnati, and flour in Minneapolis.

CHAPTER XIV

TRANSPORTATION

185. Fundamental to industry and commerce. Transportation is the one factor which can never be spared. Without it a city like New York could not exist, for its people could neither get food, nor materials for their shops and factories. Without it the Western prairies and plains could have but a scattering people, for the grain and cattle could not be marketed and manufactured products could not come in. Life would be reduced to the primitive stage. Civilization means variety of food, clothing, furniture, utensils, objects of art, and leisure to do one thing well. Exchange of products alone makes this possible. The higher life of man is therefore built on transportation. This principle is thus given by Macaulay: " Of all inventions, the alphabet and the printing press alone excepted, those inventions which abridge distance have done most for civilization."

186. The old era of waterways and earth roads. The first highways were streams followed by the small boats of the explorer and settler. The first earth roads were improved trails connecting settlements in the same colony. Then larger roads joined settlements in different colonies, until there was some sort of a road from Massachusetts to Georgia. There was a time when it took a freight wagon 115 days and cost its owner $1000 to go from Boston to Savannah.

The most important waterways of the colonial time were as follows: (1) the Hudson-Mohawk, with a carry from Fort Orange (Albany) to Schenectady, continuing by the Oswego River to Lake Ontario and Niagara or by the Seneca River to western New York; (2) the Susquehanna, from Baltimore and Chesapeake Bay to western New York; (3) the Potomac, James, and other tidal rivers of Virginia; (4) the Ohio from Pittsburg, and the lower Mississippi. In the second half of the eighteenth century, 1750 to 1800, several roads were laid out, reaching across the Appalachian highlands. Among these were the Genesee Road, continuing the road up the Mohawk into western New York; the Forbes Road, completed in the time of the French and Indian War, and continuing the Philadelphia and Lancaster turnpike westward by way of Bedford to Pittsburg;

FIG. 129. The *Mauretania* in New York bay

the Wilderness Road, continuing the road through the valley of Virginia and crossing through the Cumberland Gap to Kentucky and the Ohio River.

187. Turnpikes and canals from 1800 to 1850. The need of joining the regions east and west of the Appalachians by good highways was understood by Washington and other statesmen early in the history of the national government. If men and products could not go freely between the Atlantic coast and the Mississippi Valley, there was danger of estrangement and separation. There was an improved road built by the state of Maryland from Baltimore up the Potomac to Cumberland. In 1811 the national government began to build from that point the National Road, which ran to Brownsville on the Monongahela to Wheeling, to Columbus, and into Indiana. It was the best long road in America at that time, and was for the freights and mails of a hundred years ago what railways are to-day.

There was keen interest in canals. The Erie Canal was begun in 1817 and finished in 1825. It revolutionized New York and did much to develop the West. By 1835 there was through service from Philadelphia to Pittsburg. It was by rail to the Susquehanna River, by boat on the river and a canal to Hollidaysburg, by the portage railway (horses and stationary engines) to Johnstown, and by canal to Pittsburg. In 1828 the Chesapeake and Ohio Canal was begun at Georgetown, near Washington. It was carried westward in order to connect with the Ohio River. It was never dug beyond Cumberland, Maryland. Additional canals fed the Erie in New York and joined the Lakes to the Ohio and Mississippi in Ohio and Illinois. This was the first great epoch of canal building in America.

188. Railways and river traffic to the time of the Civil War. The period of the earth roads and canals passed quickly, because inventions brought in the steam engine and the iron road. The Baltimore and Ohio Railway was begun at Baltimore in 1828, the year of the beginning of the Chesapeake and Ohio Canal. In 1853 the first train from the East passed over this road into Wheeling, on the Ohio River. The first link of the present New York Central Railroad was between Albany and Schenectady, and the "De Witt Clinton train" first ran over it in 1831. In 1850 seven companies operated roads in the chain from Albany to Buffalo, and there was change of passengers and freight at each terminal. The Pennsylvania Railroad Company was incorporated in 1846 and ran trains from Philadelphia to Pittsburg in 1854. There were through rail connections from the Atlantic seaboard to Chicago in 1853.

Meanwhile a vast traffic grew up on the Mississippi River and the Ohio by the use of stern-wheel steamboats drawing but a few feet of water. Pittsburg, Cincinnati, Louisville, St. Louis, Memphis, and New Orleans were the centers of this passenger and freight carriage. Since the time of the Civil War, and by reason of the extension of the railroads, this traffic has declined. Twice as many steamboats arrived at wharves in St. Louis in 1886 as in 1906. Through boats between St. Louis and New Orleans are no longer in operation. Packet boats still run from St. Louis to St. Paul, to Memphis, and to Waterloo, Alabama, on the Tennessee River; also from Cincinnati to Pittsburg and Memphis. The Tennessee with its branches has 1300 miles of river suited to steamboats, but the traffic is small.

There is, however, a vast increase of barge traffic on the Ohio and Mississippi, dating from shortly after 1880. It is chiefly in coal and lumber.

189. Railway extension and consolidation from the Civil War to 1890. Soon after 1850 extensive surveys were made for roads to the Pacific coast. Construction was interrupted by the panic of 1857 and by the Civil War (1861–1865), but was active again after 1868. The greatest growth was from 1880 to 1890. In these ten years 70,000 miles of railroad were built, raising the mileage of the United States from 93,000 to 163,-000. The completion of the Union Pacific and the Central Pacific railroads in 1869 gave the first transcontinental connection from the Atlantic to the Pacific, the western terminal being San Francisco. In the same year Cornelius Vanderbilt completed the consolidation of the short lines leading from New York by the Hudson River to Buffalo.

Several systems now include the greater part of the railroads of the United States. These are often known by the names of men who have led in construction, consolidation, and management. The Vanderbilt lines extend from Boston and New York to Chicago, and many branches reach out into territory along the main lines. They include also the Lackawanna and Erie roads to Buffalo, the Lake Shore and Michigan Southern and the Michigan Central to Chicago, the Chicago and Northwestern, and others, forming a network extending to Iowa, Minnesota, and Nebraska. The Pennsylvania system includes roads once belonging to more than 200 companies, and forms a network covering New Jersey, Pennsylvania, Maryland,

Virginia, and the three states north of the Ohio River, westward to St. Louis.

West of Chicago and St. Louis the following are the greater groups : The Gould group covers the region from St. Louis to the Gulf of Mexico and westward to Salt Lake City, with eastern connections reaching Buffalo, Pittsburg, and Baltimore. The

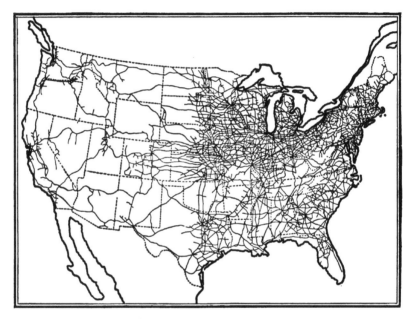

FIG. 130. Railway net of the United States : density greatest in the northeast; least in the west; intermediate in the southeast

Harriman group consists chiefly of the transcontinental Union Pacific and Southern Pacific, with extensions on the Gulf and the Pacific coast. The Hill group is largely developed in Illinois, Iowa, Nebraska, and Minnesota, and reaches the Pacific by the Northern Pacific and Great Northern railways, sending spurs northward into Canada. New England is chiefly covered by the Boston and Maine system in the north, and the New York, New Haven and Hartford system in the south. The Southern Railway system is the most extensive group south of the Potomac and Ohio rivers.

Railway building has gone on more slowly since 1890. There are now 240,000 miles of railroad in the United States, or more than two fifths of the total for all lands. Steel rails were invented about 1867. Their greater strength has meant larger cars and engines, greater speed and safety, lower rates, and a vastly extended domestic and foreign commerce. Trunk-line rates are a small fraction of what they were fifty years ago, great cities have grown, and vast fields have come into use through the development of railroads.

190. The new era of waterways. It has been found during the past ten years that the railroads, extensive as they are, can-

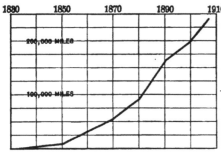

FIG. 131. Increase in railway mileage in the United States from 1832 (229 miles) to 1908 (240,839 miles)

not handle all the freight that is provided by the growth of population and the opening of farm lands in the interior. Hence has arisen a new movement for the revival and perfecting of waterways, both canals and rivers. Railway rates will be thus regulated. The old canals both in this country and Great Britain have often been absorbed or controlled by railroads to prevent compe-tition. A Pittsburg manufacturer has said that when, for any reason, navigation stopped on the Monongahela River, railroad rates rose elevenfold. Coal was carried from Pittsburg to Memphis in 1903 by barges at 42 cents per ton. In cars the cost was $3.73 per ton. The construction of the Erie Canal is said to have reduced charges to one tenth of former figures.

The supplies of iron would be conserved, for more iron is required for building railways and rolling stock than for con-structing ships. The waterways carry the bulky and cheap freights, in which speed of transportation is not important. To the railroads are left the valuable freights of small bulk and the perishable products, which must be carried quickly.

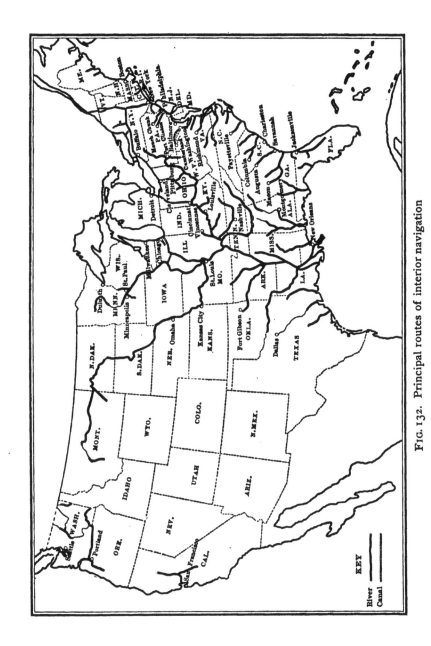

FIG. 132. Principal routes of interior navigation

KEY

River ——————

Canal ————————

235

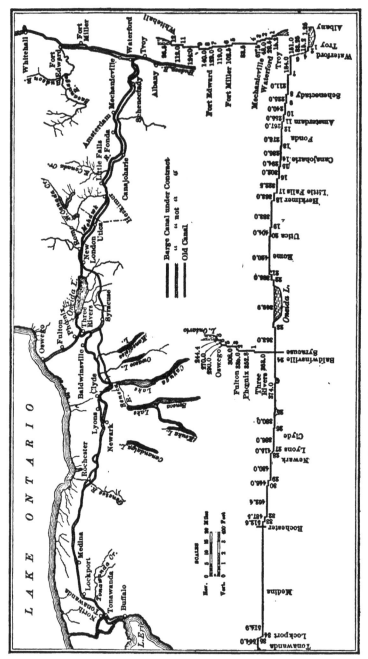

FIG. 133. The Barge-Canal route of New York

236

Of 4200 miles of the older canals more than one half has been abandoned. The chief canals now in use are in New York, Ohio, Illinois, and Louisiana. The great modern canals are usually deep, and are short links in ocean, lake, or river navigation. Thus the Suez, Corinth, and Kiel canals join parts of oceans or salt seas. The St. Marys Canal joins two of the Great Lakes. The Manchester Canal extends sea traffic to an inland city. The Panama Canal will join two oceans. The canal at Louisville joins the parts of the Ohio River which are above and below the falls. The Erie Canal is now being deepened to twelve feet to accommodate thousand-ton barges. It will thus be intermediate in character between the old and new types, and will be a link between the Lakes and the ocean. The Welland Canal is a link between Lake Erie and Lake Ontario. The Chicago Drainage Canal is suited for ships, but, to become useful, must be prolonged by the improvement of about 300 miles of navigation on the Desplaines, Illinois, and Mississippi rivers. This would extend traffic to St. Louis, and there remains the improvement of the lower Mississippi River. This is one of the largest projects now under discussion. Such a work completed would open the Great Lake and prairie country to the Gulf, and, by the Panama route, to western South America, the Pacific, and the Orient. It is claimed that these changes will regulate transcontinental freight rates in the United States. An important Canadian plan, which will affect the commerce of the United States, is to construct a canal from Georgian Bay to the tide waters of the St. Lawrence.

The Panama Canal, begun under French direction and with French capital, is now under rapid construction by the government of the United States, and it is hoped that it may be open to ships in 1915. It will create a new trunk line of ocean trade, and will no doubt be comparable in influence to the Suez Canal. As the latter diverted trade from the route around the Cape of Good Hope, so the Panama Canal will no doubt cut off most of the traffic around Cape Horn, and will bring the west coast of South and North America much nearer to the eastern United

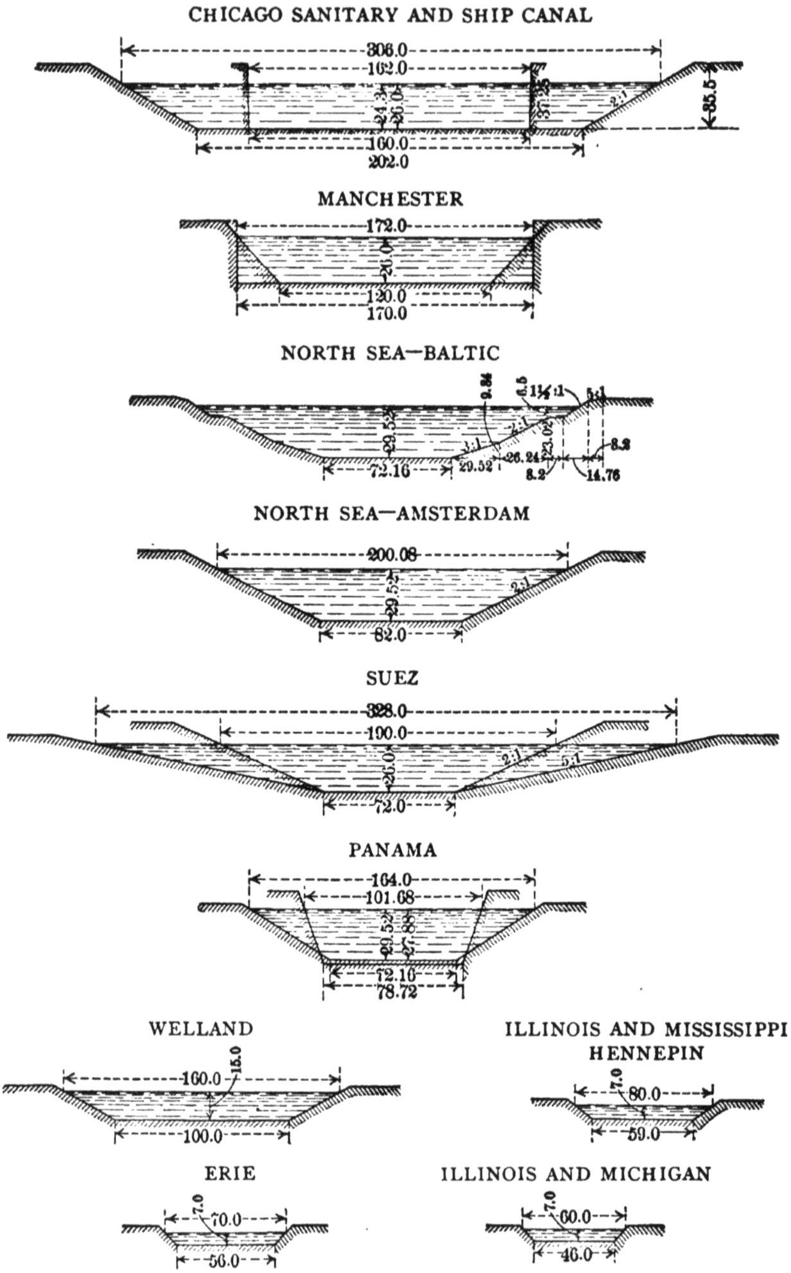

CHICAGO SANITARY AND SHIP CANAL

MANCHESTER

NORTH SEA—BALTIC

NORTH SEA—AMSTERDAM

SUEZ

PANAMA

WELLAND

ILLINOIS AND MISSISSIPPI
HENNEPIN

ERIE

ILLINOIS AND MICHIGAN

FIG. 134. Cross sections of well-known canals

States and to Europe. It is believed that transcontinental traffic in the United States will be relieved of the present burden of railway monopoly. This waterway will be a lock canal, the great Gatun dam serving to hold up the waters of the Chagres River and give lake navigation across a large part of the isthmus.

Streams numbering 295 in the United States are considered navigable, making 26,400 miles of navigation. If suitable rivers were canalized, the length would be much greater. Thus

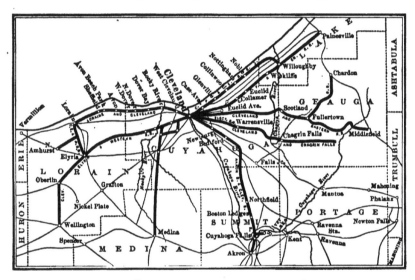

FIG. 135. Electric railways about Cleveland, Ohio

the government has placed dams and locks on the Monongahela, resulting in a vast traffic from the coal beds of West Virginia and Pennsylvania. The Mohawk will be in part canalized in the barge-canal improvement in New York. A further present difficulty with inland navigation is that rivers are of different depths and disconnected, discouraging traffic by costly transfer on land or between boats of different draft.

191. The electric railway. Electric roads have been in operation since 1889. By 1902 nearly all street railways in towns and cities had adopted this kind of power. Since 1895 interurban lines have multiplied rapidly. By electric cars one could now go with little break from Chicago to New York. The greatest development is in the Lake and prairie states from Ohio westward. Hundreds of companies have been formed, and the roads form a network joining the towns in every direction. There is great improvement in the equipment and service within a few years. The cars are larger, the roadbeds smoother, speed is greater, and there are dining, freight, express, and other special kinds of car. These roads have taken much local traffic from the steam lines, to which the long-distance travel is left. The electric lines are, however, often owned or controlled by the steam roads. Electric lines build up the suburban towns and favor the movement from city to country. To the farmer they offer convenient markets and schools, greater social opportunity, and increased valuation of property.

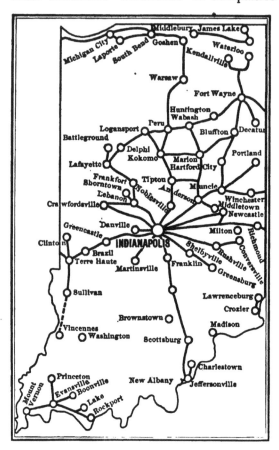

FIG. 136. Electric (interurban) railways of Indiana

192. Earth roads. The new features of transportation include not only improved waterways and electric roads but also improved earth roads. In recent years both the national government and many states have taken steps in this direction, and thousands of miles of country roads have been improved. Several things have encouraged the movement. The invention of

FIG. 137. Two animals drawing one bale of cotton over a bad road

FIG. 138. Two horses drawing twelve bales of cotton on a macadamized road

the bicycle caused many thousands of people to use their influence for good roads. A like result has come more recently from the widespread use of the automobile. In some Western states the automobile is much used between the farm and the

market. The adoption of the plan of rural free delivery by the
Post Office Department has been effective because the Department
refuses to send its carriers over bad roads. The railroads see
the value of widening the belt of farm lands from which they
may receive passengers and freight, and in some cases have sent
" good-roads trains" over their lines, bearing government experts
and suitable machinery, to build sample roads and educate the
people upon this subject. The perfecting of earth roads is a
direct benefit to all classes of people.

193. Express business in the United ·States. In the early
days the driver of a stage often transacted small items of business for people living along his route, making purchases and
delivering parcels. The express companies now do this service
on a large scale, mainly in coöperation with the railroads. In
foreign countries more of this work is done by the parcels post
and by fast freight. Independent companies are usually organized, and they pay a share of the receipts to the railways for
cars and hauling. In smaller towns the local agent of the railway serves also as express agent. The express companies also
issue money orders and letters of credit to travelers, and act as
agent in many transactions, such as filing legal documents and
passing goods through customhouses. In 1907 the express
companies issued over 14,000,000 money orders, doing more
than one fourth as much of this business as the United States
Post Office Department. Express mileage operated in 1907 was
235,903, of which but 17,795 miles were on steamboat lines and
1134 by stage. The American Express Company was organized
in 1850 and the United States and Adams companies began in
1854. The six companies owning 92.7 per cent of all express
mileage in the United States are as follows (1907):

Adams	34,862
American	42,361
Pacific	23,661
Southern	31,434
United States	30,101
Wells Fargo & Co.	43,914

194. Coast routes. The coasting trade of the United States is very great, and must, by law, be carried on wholly by American vessels. The coast line measures about 5700 miles, but if all indentations and islands are included, it amounts to more than 60,000

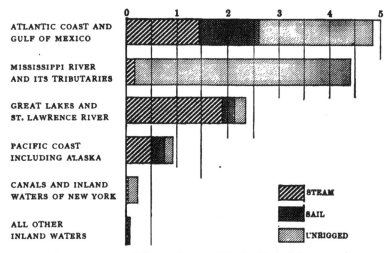

FIG. 139. Gross tonnage of all vessels owned in the United States and operated in coastal and interior waters, 1906; by geographical divisions and classified as steam, sail, and unrigged. (In millions of tons)

miles. The following statement represents, according to Professor J. Russell Smith's "Ocean Carrier," the number of companies doing business from the principal Atlantic ports in 1907:

Boston. . . . 7 companies, serving 10 ports (Eastport to Jacksonville)
New York . . 12 companies, serving 19 ports (Portland to Galveston)
Philadelphia . . 5 companies, serving 8 ports (Boston to Savannah)
Baltimore. . . 3 companies, serving 6 ports (Boston to Savannah)

Other routes originate at Mobile and New Orleans on the Gulf coast, and at San Francisco, Portland, and Puget Sound ports on the Pacific coast. Important types of ship have place in the coasting trade, such as barges for coal and steamships planned especially for the great fruit trade with the Gulf States and the West Indies.

An inside waterway is an important project for the Atlantic coast traffic. The scheme involves short canals, as across Cape

Cod, and enlargement of the present canal from the Delaware to the Chesapeake. Thus by using the canals and numerous lagoons and bays a protected water route becomes possible along most of the Atlantic sea border of the United States.

FIG. 140. Steamships in the locks at the " Soo "

195. Great Lake routes. Duluth, Minnesota, and Ogdensburg, New York, are the extreme west and east points of navigation, and the distance by the lake route between the two is 1221 miles. By reason of Niagara the heaviest traffic is confined to the four upper Lakes. The Soo Canal and the improved channels of the St. Clair River, St. Clair Lake, and the Detroit River give passage to large vessels down the upper Lakes. On account of ice, navigation is here limited to about seven months in the year.

The trunk lines on the Great Lakes have been defined as follows :

1. Chief trunk line from Duluth and other ports on Lake Superior to Toledo, Cleveland, Buffalo, and other places on Lake Erie.

2. From Lake Michigan ports — Chicago, Milwaukee, and others — to Lake Erie. This line was formerly more important

than the preceding. The change is due to recent developments in iron and grain on lands tributary to the Lake Superior route.

3. From both Lake Superior and Lake Michigan some traffic goes to ports on Georgian Bay. This is especially true of the Canadian ports, Fort William and Port Arthur.

4. Some traffic continues from the chief trunk line by the Welland Canal, Lake Ontario, and the St. Lawrence River.

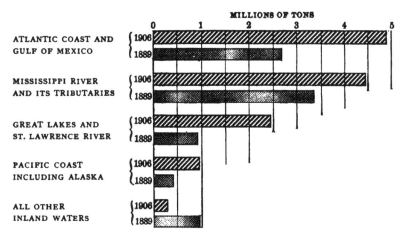

FIG. 141. Gross tonnage of all vessels by divisions, 1889 and 1906

5. Considerable trade passes between Lake Michigan and Lake Superior.

The rivers entering the Great Lakes are short, and are unimportant for navigation except as connecting links and as affording harborage. The Cuyahoga River at Cleveland and the Chicago River are examples of the latter use.

196. Ocean routes. Some of the important routes of ocean traffic are defined by Smith in the "Ocean Carrier." First is the North Atlantic trunk route, from the English Channel to New York bay. By the Channel ships reach London, Southampton, Hamburg, Bremen, Rotterdam, Antwerp, and several French ports. A slight diversion to the north leads to Liverpool, Bristol, and all Welsh ports. One sixth of all ocean shipping is said to pass over this route. The Mediterranean-Asiatic trunk route passes through the Strait of Gibraltar and around Asia to Japan. It has terminals both in the English Channel and in New York. The large passenger and other traffic from New York to Naples and Genoa follows this route in part. The route reaching from Japan and other east Asian ports to the Pacific coast of the United States may be regarded as a continuation of the Mediterranean trunk route.

The Good Hope route extends from New York and the English Channel to Delagoa Bay, Australia, and New Zealand. It is a great track for sailing vessels. The South American trunk route is likewise supplied from two sources, — New York and neighboring ports on the one hand, and the English Channel on the other. From Cape St. Roque there is a common course around South America and up its west coast, with a branch to New Zealand. Thus a ship leaving New York may go to western Europe, or to the Mediterranean, around the Cape of Good Hope, or to South America and the Pacific. The opening of the Panama Canal will work great changes in the movement of American commerce toward markets bordering the Pacific Ocean.

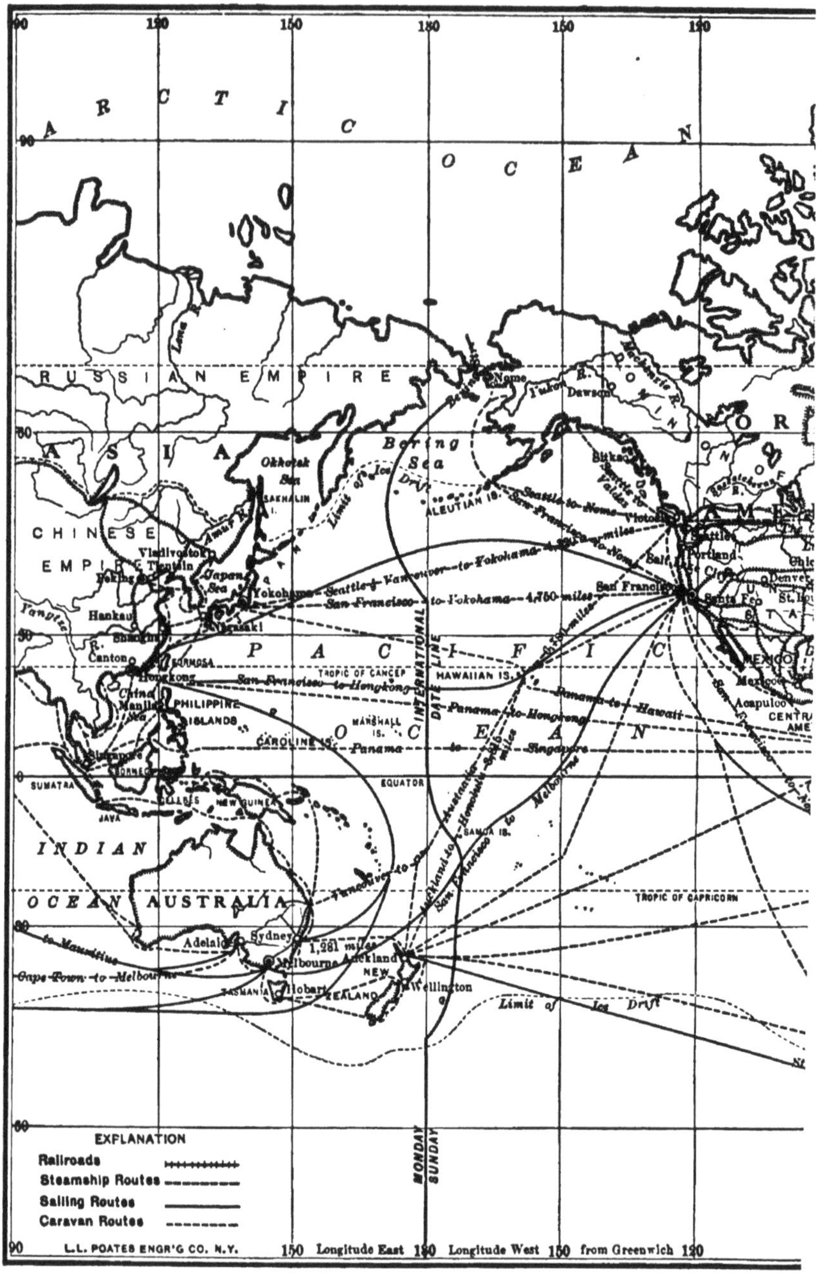

Principal transporta

GREENLA

Limit — *Drift*

ortation routes of the world

CHAPTER XV

197. Old and new ways. Before the days of telegraph and telephone a letter or messenger offered the quickest means of sending intelligence or making a bargain with a distant person. And the letter or messenger went not by fast train or steamship, but on horseback, by stage, or sailing craft. Weeks might be required for an answer on land, and months if correspondence were across the sea. The wars and politics of Europe were reported in American newspapers five or six weeks after the event. Commerce with Europe in ideas and in commodities proceeds now at no such dragging pace. Two or three weeks suffice for a return letter, a few minutes for answer by cable, and goods may come or go by fast ship in a week or less. At home several days were taken for a transaction now made instantaneously over a wire. A stock or grain market can be used daily by the whole country, and quotations at important centers like London, Liverpool, New York, or Chicago at once affect the markets of the world.

Indirectly swift communications affect commerce in many ways. A daily weather service is possible only by telegraph and telephone, by which the observations of many observers can be at once gathered, plotted on maps, and sent out to the public from many local stations. These weather reports are especially important to coastwise shipping and to the farmer. Newspaper service has been revolutionized in the same way. Press associations, with agents everywhere, gather news and sell it to the newspapers of cities and larger towns, so that it is read within a few hours after the occurrence. The modern man of business knows at once the markets, the condition of crops, the new developments in mining, new organization of capital, new government regulations, inventions, and the condition of public sentiment as affecting commercial transactions. One may seek a file of a newspaper of seventy-five years ago and observe the dates of issue as compared with the dates of occurrences. It is now possible for a commanding officer to control an entire campaign, knowing what has happened at each hour and transmitting his orders accordingly. Wireless telegraphy will give the same precision of command to naval warfare, and whether battles are on land or sea, the nations are likely to know the events within twenty-four hours. Domestic convenience, the saving of life, and relief in disaster are all secured by swift communication. At the time of the San Francisco earthquake of 1906 relief began to arrive in the city in a shorter time than would have been required in 1850 for the news to reach the cities of the Atlantic coast. When Messina was destroyed houses were being built by American money within the time once required to bring the news even to London.

198. The railway-mail service. As the railways only gradually perfected their tracks and rolling stock and thus gained regularity and speed, the mails were not at once transferred from horseback and stage to steam cars. The first ten years of this slow and broken railway service were from 1830 to 1840. Mails sometimes made as much as fifteen miles per hour by stage. Stages also ran all night, while the first railway travel was chiefly by daylight. Heavier rails, better locomotives, better bridges, and the union of short lines all helped toward better mail service. The first night mail between Boston and New York was put on in 1860. In the old days mail was all sorted at the post offices and the stages waited for the mail bags to be made up. This was at first common in rail service, but trains could not wait, and the result was that mails were left behind and arrived at their destinations much later than passengers. The essential need was the railway post office. An enterprising mail agent at St. Joseph gained permission in 1862 to go to the eastern terminus of the Hannibal and St. Joseph Railroad and sort the mails on the train, so that they might be ready for the Western stages on arrival at St. Joseph. New cars were then planned and built, and the work has grown until exclusive mail trains with all post-office conveniences now have right of way and make the fastest time. These exclusive trains were first used about 1875 between New York and Chicago, on the New

FIG. 142. Interior view of Southern Pacific steel postal car

York Central system, and soon on the Pennsylvania Railroad. Mail in some cases is sorted and ready for city carriers on arrival. Not all mail goes to a central post office, but is directly transferred from one mail train to another.

199. Classification of mail matter. Four classes are established : (1) letters and sealed packages ; (2) newspapers and other periodicals ; (3) books and other printed matter not periodical ; (4) merchandise. The rate for the last is one cent per

ounce. Merchandise is extensively carried by mail in many foreign lands. The shorter distances are there favorable to this
policy, but it is generally believed that ample parcels-post facilities would have been adopted in the United States but for the
influence of the great express companies upon legislation. The
sending of money by post-office orders is extensive and payment is guaranteed. Valuable letters and packages are registered, thereby increasing safety, with indemnity in case of loss.

FIG. 143. Rural free-delivery carrier and mail wagon, Fountain, Michigan

An additional postage of ten cents insures the special delivery of
a letter on its arrival at the post office. Official business of the
government and its representatives is " franked," or carried free.
Cities and towns whose postal receipts reach a certain amount
have a free-delivery system. As countries become densely
peopled the tendency is to establish such delivery for all.

200. Rural free delivery. The first experimental routes were
centered at three towns in West Virginia in 1896. By November, 1899, there were 634 routes, serving nearly 500,000 people. One year later, in 1900, there were 2551 routes, serving

1,800,000 persons. At the present time the service has become very general throughout the United States. It has been found to increase the postal receipts and to enrich the conditions of rural life. Some saving toward the expense is made by discontinuing small post offices thus made needless. The carrier also delivers registered mail and may act as agent for securing money orders at the central post office. The service has resulted in improvement of roads, in early breaking out of roads after winter storms, in regularity of farm supplies for towns, as well as regularity of mails. A saving of time is effected in thus organizing a few men to make the journey between the homes and the post office. There is also advantage in prompt knowledge of markets. Among many cases reported, two Kansas farmers made the annual price of their daily paper by increased profits on single loads of hogs, due to prompt information. At a post office in Oregon 13 daily papers were taken. Three years after the establishment of free-delivery service 113 were taken.

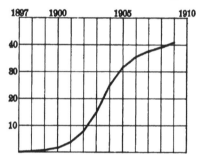

FIG. 144. Increase in the number of rural free-delivery routes in thousands, 1897–1909

201. International postal arrangements. Under international compacts made in 1874 and 1878 the International Postal Union has an office in Bern, Switzerland, and a meeting is held every three years to consider questions that arise in the transmission of mails. Thus nearly all civilized countries are a single postal territory. Much has been done for uniform and cheap postage. Five cents or its equivalent is the usual postage for a letter not above one-half ounce in weight. By a recent arrangement such letters now pass between the United States and Great Britain on a postage of two cents, a rate that has long prevailed between the United States and Canada.

202. The land telegraph. The first business line in the United States was put into operation in 1844. This kind of communication falls naturally into four subdivisions: (1) commercial telegraph for the business of the general public; (2) railroad lines for train dispatching and other business of the railroad companies; (3) government systems; (4) municipal systems. In 1907 there were 25 commercial companies in the United States, but this number includes companies owning also ocean cables. Much the largest controls in this business are held by the Western Union Telegraph Company and the Postal Telegraph-Cable Company. Over 29,000 offices were reported in 1907, of which 22,000 were in railway stations. Including cable messages, over 100,000,000 messages were sent in 1907.

Train dispatching by telegraph was first put into operation by the Erie Railroad in 1846. It would be impossible to manage the heavy traffic of trunk lines at the present time, even with double or quadruple tracks, without such means of communication. Several railroad companies now conduct schools of telegraphy, to provide a sufficient number of competent operators.

Government telegraphs are used by the Signal Corps of the United States army, by the Weather Bureau, and by the Panama Canal Commission. Municipal systems principally serve the purposes of fire alarm and police patrol.

203. The submarine telegraph. In 1902 companies in the United States owned 16,000 miles of submarine cable. The increase was very rapid during the following five years, and in 1907 the total was 46,000 miles. The chief new cables of that period were the Pacific, the New York and Havana, and the New York and Colon. Others were laid down, mainly on account of the colonial development of the United States. The Pacific cable extends from San Francisco to Hawaii, Guam, the Philippine Islands, China, and Japan. It is 8000 miles long and was laid in eighteen months. This is the greatest of single ocean cables.

Throughout the world 252,000 miles of submarine cable were in use in 1904, of which 139,000 miles belonged to English corporations and 15,000 to the British government. The British lines include five across the North Atlantic, and a Pacific cable

FIG. 145. Principal submarine telegraph cables of the world

joining Vancouver to the Fiji Islands, New Zealand, and Australia. The United States is in this respect second among the nations, and also has five lines across the Atlantic Ocean. The earliest ocean cable joined Dover and Calais across the English Channel, and was laid in 1850.

The wireless telegraph is now in large use in the mercantile marine, in the armies and navies of the world, and, in a more experimental way, on land. Overland communication is carried

FIG. 146. Wireless-telegraph station, Glace Bay, Nova Scotia

on by the government between Washington and the Brooklyn Navy Yard. A station in California heard and took down communications held between Pensacola, Florida, and the United States steamship *Connecticut*, when the latter was off the east coast of Cuba ; the distance was 2800 miles. Messages were sent in 1902 between Cape Cod and Cornwall in southwestern England. Since 1907 some commercial business has been done between stations at Cape Breton, Nova Scotia, and Clifden, Ireland. Five commercial wireless companies were reported by the census of the United States in 1907. The most notable service of wireless telegraphy to date was in summoning other ships and saving several hundred persons, the passengers and crew of the steamship *Republic,* when it suffered a collision near Nantucket in the summer of 1908. All the greater passenger steamers are now equipped with wireless apparatus, and thus upon the frequented ocean routes are never out of communication with other vessels. Wireless transmission is to be regarded as in its beginnings.

205. Alaskan and insular service. In 1907 there were in Alaska 4000 miles of land telegraph lines and in Alaskan service 2000 miles of submarine cable. The principal submarine line . runs from Seattle to Sitka. There was also a wireless system operating between stations 107 miles apart. It is proposed to arrange for wireless messages between Alaska and the mainland of the United States, also reaching the Yukon River and other parts of the interior where the maintenance of ordinary land lines would be difficult and costly. The Signal Corps of the United States army is in charge of this system.

In the Philippine Islands service is maintained partly by the Signal Corps and partly by the civil government of the Islands. In 1907 there were in the Philippines 6000 miles of land line and over 1400 miles of submarine cable. In 1907 the Porto Rican telegraph service had 774 miles of wire and transmitted over 200,000 messages. A school is organized at San Juan for the training of operators.

206. The telephone. Communication by telephone was in its infancy in 1880. At the present time the volume of business is much greater than that carried on by telegraph. The latter has been the chief means of sending messages over long distances, but the long-distance telephone has now come into large use and is of growing importance. The telephone for local communication is practically universal in country, town, and city. The city office can at once speak with thousands of business houses and private homes over a radius of many miles, and the farmer may, without leaving his home, call not only his neighbors but all towns and cities in his own and often in neighboring counties. A vast total of time is saved and business is transacted easily and swiftly. According to the census the telephone industry was three and one-half times as large as that of the telegraph in 1907, and over 8,000,000 miles of wire were added to the telephone service of the United States in the five years previous to that date. In comparison, but 259,000 miles of wire were added to the telegraph lines in the same period. The totals in 1907 were:

Telegraph wire	2,000,000 miles
Telephone wire	13,000,000 miles
Total	15,000,000 miles

This length of wire would reach 600 times around the globe at the equator.

207. The world's telegraphs. In 1907 there were 48 countries having telegraphic systems under government ownership, and these countries contained 945,000,000 of the world's population. All adhere to a code of rules adopted in St. Petersburg in 1875, and an annual report is made through an International Telegraphic Bureau located at Bern, Switzerland. Most of the private telegraphic companies of the world adhere more or less closely to the St. Petersburg agreement. Thus the world is becoming one through a growing network of almost instantaneous communication. The remote farmer, ranchman, or miner is given neighbors; and the discoveries, the business, the battles, and the catastrophies of all lands become speedily known to all civilized men.

CHAPTER XVI

GOVERNMENT AND COMMERCE

208. General objects. When the American Constitution was devised Congress was given power "to regulate commerce with foreign nations and among the several states, and with the Indian tribes." This was a broad provision, and many things are now done by the government which could not be foreseen in an early day. They belong to the spirit of the Constitution, however, and have arisen with the growth of commerce and industry. The first need is protection against robbery or other forms of dishonesty. There must also be uniform standards, so that men need not make exchanges in the dark. Each government seeks to increase the prosperity of its own country in forming its trade laws, but it is more and more found that all countries flourish or suffer together, and that world-wide coöperation is the underlying principle of commerce.

209. Weights and measures. In primitive times measures were derived from parts of the body, as the length of the foot, the length of the finger, or the width of the hand. The length of average grains of wheat or barley was also used, or the length of a pace, or the distance from the tip of the king's nose to the end of his thumb. Among modern sources of standards are a small fraction of a meridian or the length of a pendulum which beats seconds. A cubic inch of distilled water at a given temperature is often adopted as a standard of weight. Measures of contents and of weight are based on measures of length. Standards are actually embodied in metallic scales or measures of capacity, and are held in safe-keeping by the proper bureaus at national capitals. There is an International Bureau of Weights and Measures, having its headquarters in Paris; and the National Bureau of Standards, of the United States, organized in 1901, is a bureau of the Department of Commerce and Labor.

Uniformity among nations is highly desirable, since all countries enter so largely into international trade. To this end much influence has been used, that the metric system, now largely in use among European and South American peoples, may be adopted by Great Britain and the United States. In these two countries, as well as in Russia, Turkey, Egypt, and Japan, the metric system has been legalized but is not made obligatory. Error

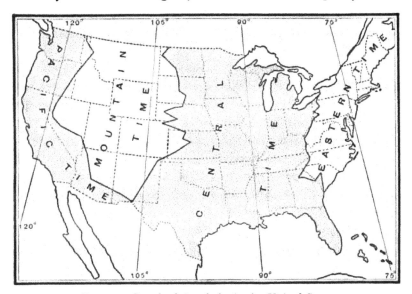

FIG. 147. Standard-time belts in the United States

or dishonesty in weighing or measuring strikes at the roots of trade, and suitable laws and penalties are of very ancient origin.

210. Time and the International Date Line. Before the days of swift transportation and instantaneous messages uniformity in time was not important. Now a train of cars often runs over so many degrees of longitude that the local time at its several stations varies several hours. There must be uniform time or delay and danger would result.

To attain this uniformity standard-time belts have been adopted, especially by the railways, so that many points on the route keep both local and railway time. Time belts were adopted in 1883 for the United States and Canada. They are reckoned from Greenwich, and are each 15 degrees wide, corresponding to an hour of time. The following names have been selected for the time of various meridians :

Colonial time	60th meridian
Eastern time	75th meridian
Central time	90th meridian
Mountain time	105th meridian
Pacific time	120th meridian

A time belt extends 7.5 degrees each way from the chosen meridians. Thus central time prevails from 82.5 degrees to 97.5 degrees. The boundaries between belts are sometimes curved, to save an important town from the need of keeping two kinds of standard time, as would happen if a city were on the margin of two time belts.

In traversing the earth to the west one goes with the sun, and more than twenty-four hours passes between the noon of one day and the noon of the next. In encircling the globe this results in the loss of an entire day. In traversing the earth to the east one meets the sun, and less than twenty-four hours passes between midday and midday. This, in encircling the globe, results in gaining one entire day, or one hour for each 15 degrees of longitude. Two persons starting from the same point and going around the earth in opposite directions, setting their watches to local time each day, would find their calendars two days apart on their return to the starting place. Hence a meridian has been chosen at which the west-going traveler will add one day to his calendar and the east-going traveler will subtract one day. This meridian is the International Date Line. It is 180 degrees from Greenwich, and was chosen because it is in the Pacific Ocean, as much removed as possible from important lands. On the principle already stated for time belts on land, the line is considered to bend eastward to avoid the neighborhood of New Zealand, and also eastward to pass through Bering Strait and thus avoid the mainland of Siberia.

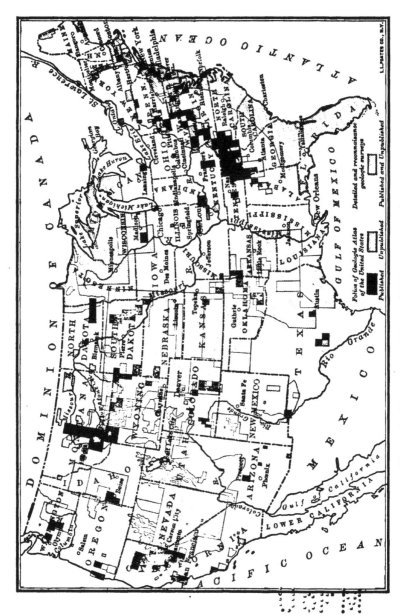

FIG. 148. Areas covered by geologic surveys to the year 1909

211. Money, banking, and insurance. Commerce can be carried on without money, by barter, and this is the primitive way. But this is a clumsy method, and trade would have remained in its infancy if some medium of exchange had not been devised. This medium is called money. Standard articles of trade have been used, such as tobacco, beaver skins, sheep, or salt. But these were awkward forms of money, useful only for a single region or a short period.

Hence gold and silver are much used because they are valuable for other purposes, are durable, easily worked, and can be made into coins of various fixed sizes. They are understood also and valued by people of all nations. Baser metals, such as copper, nickel, and iron, serve as convenient tokens of value for small sums. Paper money has no value in itself, but is given value because it stands for gold or silver and is exchangeable for them. It thus is a more convenient medium of exchange. The nearest approach to a universal purchasing medium is gold. In particular, British gold coins are accepted nearly everywhere.

In large commercial transactions, however, both in domestic and foreign commerce, actual money is little handled, drafts and bills of exchange being used. Such transfers of credit are carried on by the banks. Banks are often either government institutions or under the direct supervision and control of government. Thus our national banks issue paper money, with payment assured by the United States, and agents of the government inspect their accounts at frequent intervals. The local bank in a community not only furnishes a safe place of deposit for the surplus money of its patrons, but lends money for personal needs and business enterprise. It is thus a means of gathering up the available resources of a community and making them useful in commerce and industry. The local depositor cancels his obligations by drawing personal checks on his account at the bank, or by buying a draft at his own bank, which his creditor can collect at some other bank. The great banks of the commercial centers of the world carry on their business on a vast scale, and may finance great railways, and large manufacturing enterprises, or may even serve the needs of nations which must borrow money to pay for improvements or to meet the costs of war.

Insurance has an important relation to industry and trade. It protects against many forms of loss, being known as life, accident, health, fire, unemployment, fidelity and surety, and credit insurance. Marine insurance protects against losses at sea. The important principle is that many bear easily the burdens which would otherwise crush the few. If the owners of property had suffered the full loss in the great fires of Chicago, Baltimore, and San Francisco, disaster would have resulted not only for these cities but to a great variety of persons and industries in other places. The state governments oversee with increasing stringency the management of insurance companies.

212. The census. The Bureau of the Census is a part of the Department of Commerce and Labor. The federal census of population was first taken in 1790, manufacturing statistics were added in 1810, and the data as to agriculture began in a partial way to be gathered in 1840. Very full statistics are now collected as to population, race, conjugal condition, disease, and mortality ; also as to defective persons, crime, pauperism, education, and other matters pertaining to the study of society. Agriculture and manufactures are reported in great detail, and often with full historical treatment. Mining, fisheries, insurance, mortgages, churches, benevolence, transportation, and foreign commerce all fall within the scope of the census. The national census is taken each decade, the enumeration of 1910 being the Thirteenth Census. Many states take a census five years after each federal census. The Washington Bureau is, however, always at work in the collection and publication of special statistics relating to industries and other matters. The census is therefore a mine of information for all who are interested in any phase of industry or of home and foreign trade. Similar work is done by all civilized governments, so that a world-wide comparison is always possible.

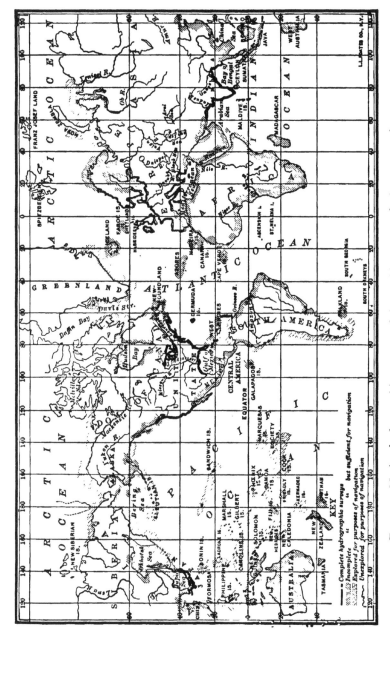

FIG. 149. Marine hydrographic surveys of the world, — state of advancement in 1904

213. Geological and geographical surveys. Since 1830 an increasing number of states have carried on geological surveys. These have dealt not only with mineral resources, but with water, soils, plant life, and general agricultural resources. Prior to 1879 several surveys were at work in the West, under various government departments. Since 1879 the present United States Geological Survey has undertaken the mapping and survey of large parts of the national domain, including Alaska. A most important product is in large-scale maps, already covering large areas, and intended ultimately to show the reliefs, drainage, and human features of the entire country. These maps serve as base maps for geological formations, mineral deposits, reclamation and forest data, and give information to settlers, travelers, prospectors, railway engineers, and the directors of industry. A soil survey and a biological survey are under the administration of the Department of Agriculture.

214. Protection of health and life. Government here acts in close relation to industry and commerce. It makes laws concerning the employer's liability for injury to the employee. Factory inspection takes account of the safety of machinery, ventilation and other matters of sanitation, and of adequate fire escapes. In many states child labor is controlled and compulsory education enforced. Hours of labor are often limited by law. Provision is made for safety in mining. By act of Congress a Bureau of Mines was, on July 1, 1910, established in the Department of the Interior. The special province and duty of this bureau is to investigate the methods of mining, especially with reference to the safety of miners and the prevention of accidents, to supervise the investigation of structural materials, to make tests of mineral fuels, and to investigate the treatment of ores and other mineral substances.

Public health is promoted by laws relating to pure food, beverages, and drugs, and laboratories are maintained for the analysis of sample products. There are laws against the pollution of water and against other nuisances. Such is the requirement that railroads in large cities shall use electric power, or the factories shall consume their smoke. The quarantine of ports and of houses or towns where there are contagious diseases belongs to the same class of government work. General sanitation on an extensive scale was carried on by the government at Havana, and is now in effect on the Isthmus of Panama, to eradicate yellow fever, typhoid fever, malaria, and other diseases. The effect on commerce may be seen in the shortening of quarantine and the more prompt forwarding of South American freight. The regulation of immigration has much to do with public health and morals as well, as it keeps out diseased persons, paupers and those likely to become such, and all criminals so far as they can be detected. The character and number of the laborers admitted to the country are also under control, as in the case of the rigid exclusion of the Chinese.

Safety in land transportation, whether by rail or other means, is an end of government. Automatic brakes and car couplings, the abolition of grade crossings, and the limitation of speed are subjects of legislation by state and municipal governments. Safety in navigation is promoted in a multitude of ways by government aid and regulation. Some of these will be named, such as the registration and inspection of steam vessels and the licensing of ship masters. Seas and lakes are charted the world over, and coast lines are studied and mapped with care and skill. Thus the Mississippi and Missouri River Commissions and the United States Lake Survey operate under the engineers of the United States army and publish an elaborate series of maps. The United States Coast Survey maps the coasts and marginal seas with soundings in detail. The United States Hydrographic Office and similar bureaus of other nations are clearing houses for the world's knowledge of all seas. The Hydrographic Office, located in Washington, publishes many hundred maps, including pilot charts of the Atlantic and Pacific, iceberg maps, derelict maps, and bottle maps, showing the movement of ocean currents.

The Weather Bureau, under the Department of Agriculture, protects commerce on land and sea by its daily weather reports

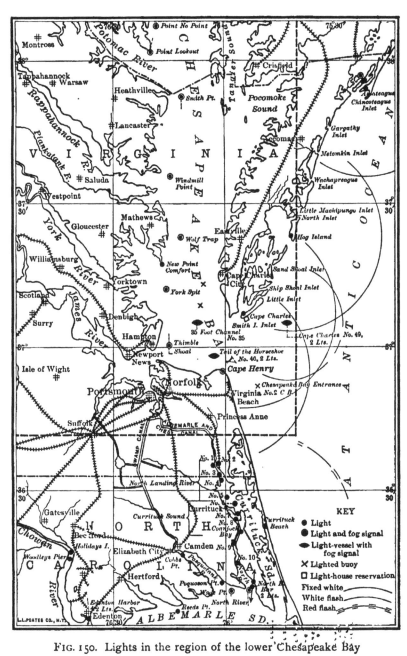

FIG. 150. Lights in the region of the lower Chesapeake Bay

and flood warnings. Predictions of storm conditions are particularly useful to the coasting traffic of the Atlantic and Gulf borders.

Lights for the guidance of ships are under the care of the Lighthouse Board, which now operates under the Department of Commerce and Labor. It is the policy of the lighthouse management that a ship near the coast should always have within sight a beacon to safeguard its course. For purposes of administration, including inspection every three months, sixteen

FIG. 151. The *Carnegie*, a nonmagnetic ship used by the Department of Terrestrial Magnetism of the Carnegie Institution

districts are arranged, including the Atlantic, Gulf, and Pacific coasts, the St. Lawrence River and Great Lakes, and the Ohio and Mississippi rivers. Four classes of lights are recognized: (1) primary coast lights; (2) secondary seacoast lights and lake-coast lights; (3) light vessels; (4) sound, bay, river, and harbor lights. Thirteen hundred fifty-four lights were enumerated in 1907. Stringent laws are enacted against false lights or tampering with official lights. The life-saving service is also highly organized and has hundreds of stations on the various coasts, at points of greatest danger. All life-saving appliances are ready for instant use, and trained surfmen are always on duty, the number being greatest from September until the following spring. Many lives, and property to the value of millions of dollars, are saved each year.

In earlier days defense against piracy was one of the chief duties of government in relation to traffic on the seas. Navies now provide for the security of commerce on the high seas when nations are at war, but international usage protects the vessels of neutral nations.

The Carnegie Institution, while not belonging to government, is located at Washington, and maintains a department of terrestrial magnetism, whose work is to make magnetic determinations and maps for land and sea, indicating the variation of the compass in many regions. There have been many expeditions in North America and Asia, and extensive work has been done on the Atlantic and Pacific oceans. The latest ocean work is by means of a new nonmagnetic ship, the *Carnegie*. Errors have been found and corrected which were sufficient to have sent ships many miles out of their course.

215. Aids to transportation. Direct assistance is given by government in several ways ; for example, in the building of roads, such as the old National Road, and many highways in European lands ; also in the construction of most canals and the improvement of rivers and harbors. In new countries land grants are often made to railway companies, which otherwise could not build long and costly lines through regions of few people. Under this head come several lines which cross the West to the Pacific coast, both in the United States and Canada. Some countries, as Italy, have put their railroads under government ownership and management. All railroads derive their charters and their right to appropriate land for their tracks from public authority. The limiting of the coasting trade to vessels of the United States is a government act in control of commerce. Of similar nature are subsidies in aid of navigation. The United Kingdom makes grants to some of its steamship lines, and holds certain rights as to the transport of mails and the use of the vessels in time of war.

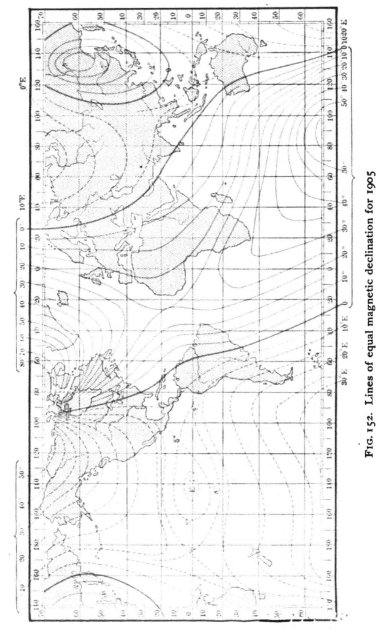

FIG. 152. Lines of equal magnetic declination for 1905

Full lines indicate west magnetic declination and broken lines east declination

216. Regulation of trade. Some illustrations of regulation have already been given, as in case of laws for purity of food and drugs. Corporations doing various kinds of business derive their authority from the state and are held by its restrictions. Such control is now especially directed toward railways, because these corporations represent immense capital, and their policy is at the foundation of trade and affects the prosperity of every citizen. In particular, government exerts its authority to prevent the payment of rebates, and other discriminations or policies which confer unequal advantage. Monopoly in any line of trade and the arbitrary fixing of prices to the detriment of the consumer have become subjects of frequent legislative and judicial inquiry.

Chief among the agencies of regulation is the Interstate Commerce Commission. Such a body became necessary as railways extended. Since a single line or system crosses many states and is therefore subject to no single state government, federal supervision is required. The first important provision was made in 1887. There has been supplementary legislation down to the present, with increased public authority over transportation. The general objects are the prevention of pooling and of rebates and all secret discrimination, the fixing of just rates, and the requirement of safety appliances. The findings of the commission are subject to revision by the courts.

Similar commissions have been appointed in most of the states. Exceptions exist in the West, where the need of transportation presses heavily; and in the East, where some legislatures are said to be controlled by the railways. New York has two Public Service Commissions, one for Greater New York and one for the rest of the state. These may be taken as representatives of state commissions. All steam and electric roads, telegraphs, and telephones are under their supervision. They determine whether cause exists for new lines, supervise the issue of securities, and have authority over rates, safety appliances, and the meeting of the public convenience in time schedules and accommodations ; subject, like the national commission, to the review of the courts.

The sale of liquors is subject to a great variety of legislation and control by the several states. The conservation of natural resources assumes importance with growing population, and the United States is following in the wake of the older nations of Europe in undertaking the ownership and control of natural wealth. Such ownership thus far relates more to forests than to mineral areas, but minerals and water powers are coming to

FIG. 153. Agricultural experiment stations in the United States

be recognized as public utilities in whose use and enjoyment all citizens should be protected from monopoly and private greed.

With each generation usage among civilized nations becomes more and more settled and crystallizes into what is known as international law. International conferences and arbitration tribunals are common in all important questions of difference or of mutual interest.

217. Aids to production. The agents of the Department of Agriculture, who are seeking new fruits, grains, and fibers in various parts of the world, are opening opportunities for the American farmer. Of similar use is a new breed of draft horses or of sheep, whose value is thus shown at public expense. The testing, breeding, and improvement of plants and animals is carried on at many federal and state experiment stations (figs. 153–154). The suppression of animal disease and the checking

FIG. 154. Barley, rye, buckwheat, and flax raised at the United States Experiment Station, Sitka, Alaska

of insect pests need only be mentioned, as also the regulation of fisheries, the maintenance of hatcheries, the stocking of streams and lakes, and the regulation and patrol of oyster fields. Here also we may add government protection of the interests of inventors, through the United States Patent Office; and also national and international copyrights, by which the rights of literary production are secured.

Permanent museums and temporary expositions bring together the commercial products and resources of all nations, and thus by education and by personal association trade is favored and extended. Such exhibitions are often organized by government or are aided by grants of money and by the coöperation of government experts and government departments.

218. Duties and internal revenue. Perhaps no government act affects the geography of commerce so much as the laying of taxes on exports, imports, and articles of domestic trade, for the currents of trade are often turned away from channels which would otherwise be followed. Taxes on exports and imports are known as duties, and are collected at various customhouses, which are chiefly but not always on the frontier. Important goods may be taken " in bond " to certain interior points, and there, on payment of duties, be delivered to the importer. Places where such goods are stored are known as bonded ware-houses. A schedule of duties is known as a tariff, and reference to a tariff is commonly made by the year of its enactment by Congress, or, more commonly, by a representative name, as the McKinley, Dingley, or Payne tariff. Tariffs are known as pro-tection or as revenue tariffs, according as their chief purpose is to foster home industry or to provide by indirect taxation for the expenses of government.

The heaviest duties are supposed to be collected upon articles of luxury and objects desired by people of wealth ; but, on the other hand, it is claimed that the necessities of ordinary living are often taxed, and that monopolies flourish behind a tariff wall. The United States taxes hundreds of import commodities and stands for a high degree of protection. Great Britain taxes a few objects, for revenue only, and is in effect the representative of free-trade nations.

Liquors and tobacco are the chief products upon which internal-revenue taxes are collected in the United States. These articles are selected because their ordinary use is not to be regarded as a necessity, and restriction would not work public harm. In time of war many other things may be subjected to internal taxation. Examples in the recent war with Spain are found in commercial paper and proprietary (patent) medicines.

219. Consular service. A consul is, strictly speaking, a commercial representative of his home government in a foreign land. A diplomatic representative, such as an ambassador or minister, usually resides at the capital of the country to which he is sent, and deals with political questions that arise between the two nations. No hard-and-fast line can be drawn, for a diplomat often has to do with matters of commerce, and a consul often must step outside the field of trade. The consul is, in practice, often an adviser and helper of his fellow countrymen, especially in the centers of tourist travel, such as Paris, London, Dresden, Geneva, Florence, Rome, or Naples.

The consular service of the United States was formerly very imperfect, because consuls were given places as political rewards and were removed with a change of administration at Washington. Much progress has now been made toward an efficient and permanent service, the policy of promotion being followed. More and more, merit and experience prevail, and many consuls are familiar with foreign languages. Because there is opportunity for a permanent and dignified occupation, young men of ability and training are likely to seek the consular service and follow it as a profession. In some of the higher schools, also, courses of study are provided to give suitable training for government service in foreign lands, whether as diplomatists or in the consular department. In such courses emphasis is placed upon the modern languages, economics, commercial geography, and international law.

Among other duties consuls assist in carrying out the tariff laws. They also seek information concerning local agriculture, mining, manufacturing, and the commercial conditions of the city and district to which they are assigned, and make frequent reports to the Consular Bureau of the State Department, by which annual, monthly, or more frequent reports are issued and distributed to interested persons at home. It is particularly the duty of the consul to consider all matters of interest to American trade, such as the conditions of markets and the crops, and all possible openings for the sale of American goods. Thus a recent monthly *Commerce and Trade Report*, picked at random, has, among many other topics, information on the following: diversity of purchases of foreign goods at Rangoon, Burma; business opportunities in Honduras; demand for fountain pens in British India; procedure for introducing American goods in Syria; trade possibilities in the Sinú Basin of Colombia; American steam pumps in Hamburg; investments in Cuba; why American brands of cement are not imported by Brazilian consumers; official estimate of the nitrate-of-soda supply in Chile; Bristol's storage facilities; mosquito extermination in Germany.

Each consul works for his own land, but the consular system, like modern trains, ships, and telegraphs, promotes the unity and progress of the world.

CHAPTER XVII

220. Foreign and home trade. The foreign trade, great as it is, is but a small part of the total commerce. There are nearly fifty states and territories, the country is vast in size, and its products, owing to differences of soil, climate, and history, are diverse. Hence there is much occasion for internal commerce. Cotton goods are distributed from New England, iron from Pittsburg, flour from Minneapolis, rice from Louisiana, fruit from California, and lumber from Washington. This trade is like that of Europe as a whole, rather than the trade of a single European nation. The chief differences are that we have a smaller population than Europe, and that there are no trade restrictions among states as there are among nations. In products Maine and California are as different as Norway and Italy.

221. Growth of foreign commerce. American commerce was small during the first half of the last century. The energies of a young people were absorbed in subduing the land. The following statement shows the totals at certain times. Both imports and exports are included.

TOTAL COMMERCE, 1800–1908

1800	$31,000,000
1850	317,000,000
1900	2,244,000,000
1905	2,778,000,000
1908	3,029,000,000

The figures for 1908 place the United States third among nations in amount of foreign trade, the United Kingdom and Germany being first and second. The United States was the fourth in exports in 1880, but since the beginning of the twentieth century has surpassed all others in this respect. This is largely due to the resources of the United States in raw materials which are required by all civilized people. These are iron, coal, copper, cereals, wood, wool, and cotton. The United States is first in the production of most of these.

The growth in manufactures has been great, especially during the past quarter of a century, and exports of manufactured goods have increased accordingly. As the country has grown older, power has been de-

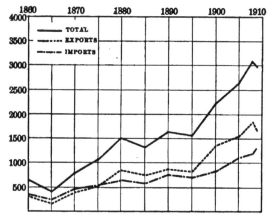

veloped, machinery has been invented, and labor has gained skill. Thus a surplus has been made in the mills and factories, which finds a larger and larger market among the nations. Exports of manufactured goods reached the amount of $23,000,000 in 1850, and of $70,-

FIG. 155. Growth of foreign commerce of the United States in millions of dollars, in five-year periods, 1860–1905, and for 1908 and 1909

000,000 in 1870; they now average more than $500,000,000 per year. If partially manufactured products are included, the last figure would be much greater.

222. Character of exports. As this is still a comparatively new country, raw materials lead in exports. Such are foodstuffs, as wheat or cattle, raw cotton, and crude petroleum. But manufactures are growing both absolutely and in percentage of exports, of which they constituted 15 per cent in 1880 and 40 per cent in 1906. The export of manufactures doubled in the ten years preceding 1906. Foodstuffs in 1890 made 42 per cent, and in 1906 only 31 per cent, of the total exports. In that year the United States was third among nations in export of manufactures.

EXPORTS OF MANUFACTURES IN 1906

United Kingdom	$1,400,000,000
Germany	1,000,000,000
United States	700,000,000
France	500,000,000

Cotton is the largest item of exports, followed by breadstuffs, meat and dairy products, iron and steel, mineral oils, and copper. Each of these exceeded $100,000,000 in value for the year 1907–1908. The export of copper has increased many times since 1890, because abundantly found, and in great demand in the recent electrical development. Another typical export is agricultural machinery. Here the American has been successful in invention, and exports have increased sixfold since 1890. The progress of the country is seen in machinery of high grade, which now enters into export trade. The list includes clocks and watches, printing presses, sewing machines, typewriters, electrical machinery, and scientific instruments, the value of the latter rising from $1,500,000 in 1890 to $11,000,000 in 1906. As population grows and skill increases, more of the foodstuffs and other crude products will be needed at home, and more manufactured goods will seek the markets of the world.

223. Character of imports. Three chief classes may be named:

1. The United States belongs to the temperate belt. Its subtropical regions, as Florida and southern California, produce but a small part of those things which we must obtain from warm countries. Hence the United States each year imports tropical and subtropical products to the value of about $500,000,000.

IMPORTS FROM TROPICAL REGIONS

Sugar	$75,000,000 to $100,000,000
Coffee	75,000,000 to 100,000,000
Rubber	40,000,000 to 50,000,000
Fibers	40,000,000
Fruits, nuts, and spices	35,000,000
Tea	15,000,000 to 18,000,000

Other items are tobacco, hides, cabinet and dye woods, gums, and silks.

2. Manufactured products of high grade, chiefly from Europe, make another class of imports. Chemicals, drugs, and dyes are a large item, particularly from Germany. Manufactured cottons, woolens, and silks figure largely, though cottons and silks of high grade are more and more made in this country. Fine glass, china,

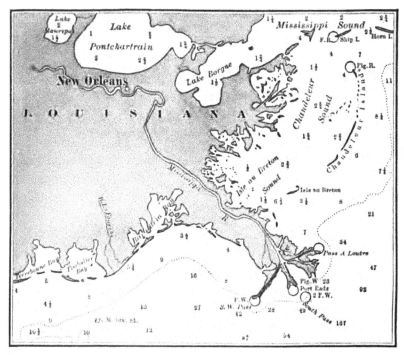

FIG. 156. New Orleans and the Mississippi delta (United States Coast Survey)

jewelry, books, maps, engravings, and the finer manufactures of wood are obtained from Europe, where labor is often cheap and skill has been acquired through centuries of practice.

3. A further class of imports is found in those mineral substances in which this country is lacking. Such is tin, amounting in value to about $25,000,000 per year. Diamonds and other precious stones make a large item of imports when times are prosperous.

224. Commercial ports of the United States. The streams of trade flow out and in at a number of ports from which radiate the trunk lines of railway and the ocean routes. On the north and south there are also long lines of land frontier with their customhouses. Most of the foreign trade, however, is by water. The Atlantic coast has now about two thirds of our foreign trade, the Gulf coast about one seventh, the Pacific coast about one seventeenth, and the northern boundary about one tenth. For the financial year 1907–1908 the percentage of foreign trade passing through seven leading ports was as follows :

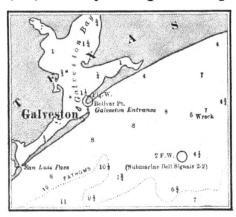

New York. . 45.48 %
Boston . . . 6.21 %
Philadelphia . 5.65 %
Baltimore . . 3.91 %
New Orleans . 6.62 %
Galveston . . 5.47 %
San Francisco 2.50 %

Seattle, Tacoma, and Portland are also important ports on the Pacific coast ;
Mobile, Pensacola, Tam-

FIG. 157. Galveston and its harbor (United States Coast Survey)

pa, and Key West on the Gulf coast ; and Savannah, Charleston, Newport News, and Portland on the Atlantic coast. Some of these smaller ports have their foreign trade chiefly with the countries of the Caribbean region.

225. Balance of trade. This is said to be in favor of the nation which sells more than it buys. In recent years the balance has been about $500,000,000 in favor of the United States, but rose to $637,000,000 in 1901 and $640,000,000 in 1908. Imports were usually of more value than exports down to 1876, but since that time the balance has in most years been in our favor. This balance is not entirely an addition to national wealth and it must not be taken as an accurate measure of national prosperity, for over against it may be set such expenditures as the cost of freight in foreign ships and the interest on much capital borrowed from Europeans.

226. Commerce with territories and colonies. Alaska has its chief industries in salmon and seal fishing and in mining. The annual catch of salmon is worth over $8,000,000. More than $18,000,000 in gold was mined in 1907, and in copper nearly $1,500,000. In the year 1907–1908 Alaska took $16,000,000 in general merchandise from the United States and sent back

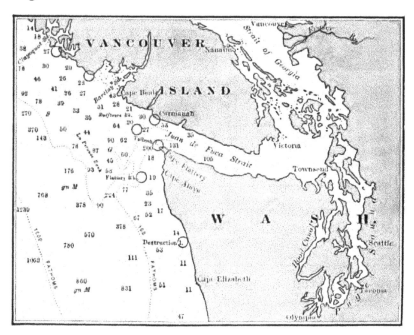

FIG. 158. Ports of Puget Sound region (United States Coast Survey)

products amounting to nearly $11,000,000 in value besides gold. There are important beds of coal and considerable timber lands, including nearly 12,000,000 acres of national forest.

The Hawaiian Islands are smaller than Massachusetts and contain about 200,000 people. The export trade, however, shows values between $30,000,000 and $40,000,000, a thirtyfold gain during thirty years of American influence and investment of capital. The islands were annexed to the United States in 1898 and organized as a territory in 1900. In the year 1907–1908 they imported from the United States merchandise to the value of $15,000,000 and sent to this country exports having a value of $46,000,000, of which sugar covered $40,000,000.

Porto Rico reported, in 1906, $23,000,000 in trade with the United States. Both import and export trade with this country increased tenfold in ten years of American rule following the Spanish War. Sugar, coffee, tobacco, citrous fruits (oranges, lemons, etc.), pineapples, bananas, and vegetables are the chief commodities. In 1905 vegetables, $4,000,000 in value, were sent to the United States. Thus the winter markets on the North Atlantic coast of the United States can be supplied with vegetables within one week from the time of gathering in the fields.

The Philippine Islands are eighteen times as great as Hawaii and thirty-two times as great as Porto Rico, and have 8,000,000 people. Their exports in recent years have averaged only about $33,000,000 per year. The exports for Hawaii and Porto Rico already stated should be put in comparison. It is further pointed out by the Bureau of Statistics that Java and Sumatra, in the same part of the world, but less than half as large as the Philippines, supported 30,000,000 people and had surplus products for export valued at $100,000,000. The chief item in Philippine exports is hemp, followed by sugar, coffee, and tobacco. Little else is sent out, yet there remains a vast field for production and the United States offers a ready market.

Transportation is the great need. Roads and ships are fundamental to commerce. When these are provided in the Philippine Islands more sugar can be raised, coffee culture revived, and it is believed rubber can be largely grown. All these are in heavy demand in the United States. There are large forest resources, and minerals, lignite, iron, and gold.

227. Other countries of North America. The United States exported to Canada in 1907 merchandise to the value of about $155,000,000. Iron and steel give the largest total, Indian corn, swine products, copper, cotton, and coal being important. From Canada in the same year merchandise worth $79,000,000 was received. Forest products, fish, and minerals supplied the greater items. Commercial relations with Canada may be expected to grow, since the common frontier is long, the people are kindred, and the resources considerably different. Northwestern Canada is continuous with the interior plain of the United States, and is far from the mother country. Many citizens of the United States are moving to the Canadian prairies, thus making strong social and commercial ties between the two countries.

On the south Cuba is the most important of the island countries, and, excepting Porto Rico, is in closest relation to the

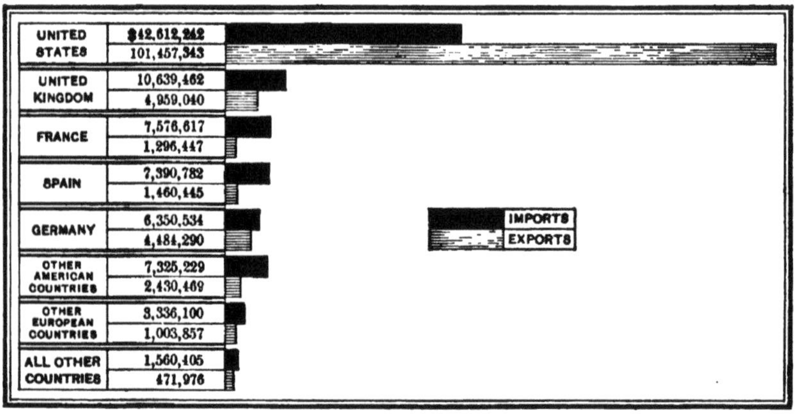

UNITED STATES	$42,612,242	
	101,457,343	
UNITED KINGDOM	10,639,462	
	4,959,040	
FRANCE	7,576,617	
	1,296,447	
SPAIN	7,390,782	
	1,460,445	
GERMANY	6,350,534	IMPORTS
	4,484,290	EXPORTS
OTHER AMERICAN COUNTRIES	7,325,229	
	2,430,469	
OTHER EUROPEAN COUNTRIES	3,336,100	
	1,003,857	
ALL OTHER COUNTRIES	1,560,405	
	471,976	

FIG. 159. Foreign trade of Cuba, — imports and exports by countries of origin and destination, — for fiscal year 1908–1909

United States. The total foreign trade of this small republic amounts to more than $200,000,000 per year, owing to the richness of its resources. This trade is about equally divided between imports and exports. The United States sends about half the imports and takes nearly all the exports, of which sugar and molasses made more than half in 1907, and tobacco made one fourth. It will thus be seen that this country receives most of its sugar from Hawaii and Cuba.

The United States holds about two thirds of both the import and export trade of Mexico. One of the largest items is American machinery, for which there is demand because Mexico has large resources and is now beginning to develop them.

Curiously, most of the cotton goods sold in Mexico are made in England, Germany, and France. The cotton goes, it may be, from Texas, just over the Rio Grande, across the Atlantic, is manufactured, and brought back to Mexican markets.

Central American countries are small, often ill governed, and rich in tropical products, but poorly developed. The United States has a considerable trade with all of these countries, and it may be expected to grow, owing to new ship lines and to American enterprise on the Isthmus. Coffee, tropical fruits, particularly bananas, and mahogany, are the leading products of these regions at present. The United States in recent years has had about half of the foreign trade of these five republics. The banana trade is also shared by the West Indian islands. Thus Jamaica, under British rule, sends to the United States nearly all of its crop of bananas, which amounts to over $5,000,000.

228. South America. The republics and European dependencies of South America, with Central America and Mexico, are often called Latin America, because settled and controlled by people of Latin origin, now speaking Spanish, Portuguese, and French. From Mexico to Cape Horn are found twenty Latin-American republics, and these, with the United States, maintain the Pan-American Union, whose seat is in Washington. Its governing board is made up of the diplomatic representatives of the republics in Washington, with the Secretary of State as chairman. Its objects are to promote friendly international relations and foster trade.

Latin America has a foreign trade of about $1,800,000,000. Of this total the United States has about $500,000,000, but the balance of trade is much in favor of the southern republics. Brazil may be taken as an example. In 1907 our imports from Brazil were valued at $85,000,000 and our exports to that country only reached $21,000,000. This means that the money received from the United States is paid to England and Germany for their manufactured products. Even France and Argentina send to Brazil almost as much merchandise as the United States. Several reasons are given for this one-sided trade. Brazil is Portuguese in language, and by tradition is in closer relations to Europe. The only fast steamship lines from the North Atlantic are from European ports. Freight ships are likely to make a triangular voyage. They load at Brazilian ports with coffee and other products for New York. There they take on bulky freights for England, Germany, or other parts of western Europe. Having discharged these, they are ready to load with manufactured products from European mills, to return to South America. Still further, American business men have usually tried to do business in South America without acquiring one of the native languages. There is therefore need of better transportation, education in Spanish, in Portuguese, or in French, and study of South American character and methods. The resources of that continent are vast,

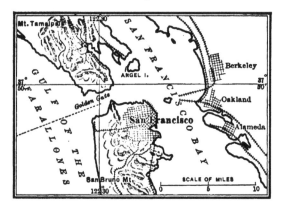

FIG. 160. San Francisco and San Francisco Bay

and our foreign trade is gaining there in spite of the difficulties named and the strong pressure of European competition.

On the west coast of South America conditions are becoming favorable to the trade of the United States. The several republics which border the Pacific and stretch back into the Andes have much agricultural and mineral wealth. They are not manufacturing countries, and the opening of the Panama Canal will give American goods the advantage of a short line. Already the sanitary improvements at Panama prevent long quarantine and hasten the transit of goods by ship and by the Panama Railroad. Much American capital is now developing railroads and mines in the countries of the west coast, and the two western continents are coming into closer relations.

Our chief exports to the South American countries include iron and steel products, such as steel rails, locomotives, and cars, and a great variety of agricultural and other machinery ; also wheat flour, cotton, leather, lard, oil, and lumber. The American Pacific coast has largely the trade in flour and lumber. Coffee and rubber are the largest imports from South America.

229. United Kingdom. Great Britain and Ireland contain about 43,000,000 people. The larger share of their food and of the materials of manufacture must be drawn from abroad. The United States produces much that the United Kingdom needs, and transportation across the North Atlantic has become the largest and swiftest offered on any ocean route. No other two nations trade together so largely. Great Britain is by far the greatest market of the United States, having taken her products to the extent of $582,000,000 in 1906 and $650,000,000 in 1907. In 1907 also the United States took from Great Britain goods to the value of $282,000,000. Foodstuffs and cottons are much the largest items in the exports from this country to Great Britain. Cotton exports were valued at $228,000,000 in 1907, or 70 per cent of British imports of this product.

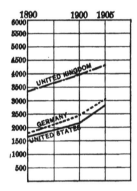

FIG. 161. Total commerce in commodities of domestic origin and consumption, of the United States, Germany, and the United Kingdom (1890, 1900, 1905), in millions of dollars

The butter and cheese trade has been lost to the United States, partly because of fraudulent brands placed on the English market, and more recently because of growing demand at home. Canada has taken a large share of this trade with the mother country. Cotton, linen, woolen, and iron goods are the leading exports from Great Britain to the United States.

230. Germany. This country is second in the foreign trade of the United States. One seventh of our exports go thither and about one ninth of our imports come from that country. Recent years have seen much growth in German-American trade. From

1897 to 1907 imports from Germany rose from $98,000,000 to $161,000,000, and exports to Germany from $136,000,000 to $274,000,000. As in the case of England, raw cotton is our largest export to Germany, giving a value of $135,000,000 in 1908. We also send considerable amounts of copper, bread-stuffs, lard, and oils. Chemicals, drugs, and dyes are large and characteristic imports from Germany, and point to the high scientific attainment and technical skill of that people.

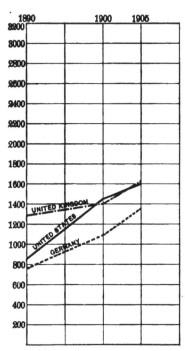

FIG. 162. Exports of domestic merchandise in millions of dollars

231. Other countries of Europe. Exports to France amounted to $122,000,000 in 1907. Raw cotton made nearly one half, and copper, machinery, and petroleum were next in value. The characteristic returns from France are in products requiring the high degree of skill and taste which belong to the advanced classes of that nation. Silk goods, fine glass, garments, feathers, artificial flowers, novelties, and works of art may be enumerated. Large numbers of Americans visit or live in France, and it is estimated that more than $100,000,000 is annually spent by these people. This does not appear in statistics of trade, but is an important element in international expenditure.

The greater part of our exports to Spain is raw cotton, again emphasizing the vast importance of this product of southern farms. From Spain come cork, fruits, wines, nuts, olives, and olive oil. American exports to Italy in 1907 were valued at $72,000,000, of which $38,000,000 were credited to raw cotton. Only Great Britain and Germany sent more goods to Italy than we, and only Switzerland and Germany took more of her merchandise. Switzerland in 1907 sent $30,000,000 in merchandise to the United States. A surprising item in this is embroideries, to a value of $17,000,000, one of the specialties of a small nation. Silks, cheese, and watches are also to be named. Silk importation from Switzerland has decreased, owing to the growth of this industry in the United States. Our exports to Switzerland nearly all go through French, Belgian, and German ports, and hence are credited to those countries.

Thus two thirds of the imports of the Netherlands from the United States are only on their way to Germany, Switzerland, and other parts of central Europe. A special feature of Netherlands trade with this country is in cut diamonds. In 1907 this trade ran low because financial conditions in America were disturbed. Diamonds, as a luxury, soon felt the loss of buying power on the part of Americans. The leading product going from the United States to the Netherlands is tobacco.

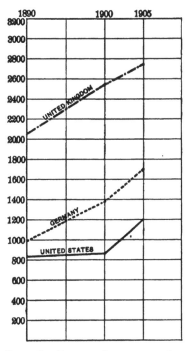

FIG. 163. Imports for consumption in millions of dollars

Of the three northern lands, Norway, Sweden, and Denmark, our trade with Denmark is much the greatest, our exports to that country being valued at $22,000,000 in 1907. Oil cake and oil meal for the dairy were the leading item ($7,000,000), and corn was sent to the value of $3,000,000. The largest item of Norwegian exports to the United States is a characteristic product, wood pulp. Turning to Austria-Hungary, we again meet our greatest export, raw cotton, of more than $30,000,000 in value. Even Russia received $12,000,000 in raw cotton from us in

1907, — one third of her importation, notwithstanding her relative nearness to Egypt and India. American agricultural implements are largely sold in Russia, as in several other European countries, though German and other imitations of our products are now competing.

In general Europe offers a market for the peculiar devices of American ingenuity, of which examples are found in agricultural machinery, sewing machines, and typewriters. American footwear has also gained a large place in the appreciation of the European consumer.

232. Specialism of countries. The trade relations of other parts of the eastern hemisphere with the United States will be noted in the study of countries. Enough examples have now been given to illustrate the basal principle of specialism as essential to commerce. A country produces what it best can, with its climate, soil, minerals, power, skill, and traditions. It can exchange what it makes with other countries of special production, by means of suitable transportation, and the process of trade is greatly helped by modern swift communication. Fine handicrafts and advanced manufactures are usually found in the older countries, and agricultural and mineral raw products prevail in the new. Thus constant adjustment and change take place, and these are affected by new discoveries, new inventions, and by the immigration of people, with consequent changes in capital and labor. Not only natural conditions, but government restrictions and tariffs, shape industries everywhere. The ideal points to perfect adaptation, perfect transportation, and unfettered exchange.

233. The United States as a self-sufficient nation. The variety of resources in the United States suggests that it might become almost a self-sustaining country. The foodstuffs now imported are largely tropical. It is believed that much of the fruit might be raised in the Gulf region and in the Southwest, while the rest might come from such dependencies as Porto Rico and the Philippine Islands. It is also believed that if the labor problem could be solved, tea could be made a domestic product. The capacity of the Philippines for coffee growing has already been mentioned. Sugar is largely imported, but the beet-sugar industry is growing and might take the large place in the United States which it has in Germany. There also seem to be no natural conditions which prevent the raising of silk. The most important mineral in which we are deficient is tin, and our temperate-latitude food products are abundant. It is not urged that self-support in this sense is desirable. When our own resources have been worked to their full measure, there will remain products which can be better or more cheaply made by others, and in the high fields of literature and the fine arts no nation could desire to be shut within itself.

The table on page 286 shows the changes that have taken place in recent years in our dependence on foreign countries for some of the most important commodities. Many show marked increase of importations. Among these are breadstuffs, coffee, diamonds and other gems, India rubber, meat and dairy products, oils, paper, raw silk, tobacco, vegetables, wood, and raw wool. The increases are partly due to growth of population and partly to greater wealth and purchasing power.

Unmanufactured fibers show little increase, while there has been great growth of importation of manufactured fiber products. On the other hand, raw silk shows a heavy increase and silk products exhibit a slight decrease, marking the development of the silk industry in the United States. Wood affords an example of large growth in importation both of raw and finished products. Wool is like silk in increased purchases of raw material, while imports of woolen manufactures have grown but slightly. Hydraulic cements, including Portland, exhibit, in 1909, less than one fourth of the importations of 1900. Inexhaustible raw materials and the establishment of many plants have nearly stopped the demand for the foreign product. Glass importations have increased but slightly, as have those of leather and of iron and steel. The heaviest imports on the list are those of sugar, and these have fallen off somewhat, a change due, no doubt, to the growth of the beet-sugar industry at home.

Increased purchases of vegetables and of manufactured cotton seem to be a needless concession to the interests of the foreign producer. As a whole the facts do not show an increasing measure of self-sufficiency on the part of the United States.

COMPARISON OF IMPORTANT IMPORTS IN 1900 AND 1909

	1900	1909
Breadstuffs	$1,803,729	$9,454,414
Cement, hydraulic	3,270,916	712,628
Chemicals, drugs, and dyes . . .	53,705,152	78,378,634
Cocoa	5,970,844	15,222,523
Coffee	52,467,943	79,112,129
Cotton, manufactures of	41,296,239	62,010,286
Diamonds and other gems . . .	14,859,018	29,373,070
Fibers, unmanufactured	26,337,805	29,748,353
Fibers, manufactured	31,152,363	49,312,392
Fish	7,472,057	12,403,012
Fruits and nuts	19,263,592	31,110,683
Glass	5,037,931	5,262,190
Hides and skins other than furs . .	19,408,217	23,795,602
India rubber, unmanufactured . .	33,041,928	64,710,370
Iron and steel	20,478,728	22,439,787
Leather and manufactures of . . .	13,292,196	13,933,134
Meat and dairy products	3,028,216	9,121,804
Oils	6,817,780	20,458,940
Paper and manufactures of . . .	3,795,645	11,632,571
Silk, unmanufactured	45,329,760	79,903,586
Silk, manufactures of	31,129,017	30,718,582
Spirits, wine, and malt liquors . .	12,758,582	23,168,845
Sugar	100,250,974	96,554,998
Tobacco, leaf	13,297,223	25,400,919
Vegetables	2,935,077	12,999,797
Wood and manufactures of . . .	14,635,340	32,351,665
Wool, unmanufactured	20,260,936	45,171,994
Wool, manufactures of	16,164,446	18,102,461

PART III. COMMERCIAL GEOGRAPHY
OF OTHER COUNTRIES

CHAPTER XVIII

CANADA

234. Chief regions. Canada and the United States alike span the continent and face the chief nations of Europe and Asia. On the east both have an indented shore, with many harbors. Tidewaters enter more deeply into Canada, but this advantage is offset by the freezing of the St. Lawrence in the winter. In the west the Canadian shore line is more broken, and, as the Pacific climate is mild, harbors are available far to the north. Canada has the advantage of short degrees of longitude in her more northern position, and thus offers a shorter route between Europe and the Orient. The land is not nearly so arctic as many suppose. It is limited, however, to products of temperate latitudes, while the United States raises much that belongs to subtropical regions. This difference must always affect the import trade of Canada.

Canada may be divided into five regions.

1. The Archæan wilderness, a region of ancient rocks and much-worn mountains, taking in Labrador, northern Quebec, much of Ontario north of the Great Lakes, and a broad belt west of Hudson Bay, running northward and including the islands of the arctic seas. Mineral deposits, furs, and forests are nearly all that is of interest to commerce, except water powers, in this region, and all of these but minerals are absent when the barren lands of the far north are reached.

2. The lowlands and low mountains of the Maritime Provinces, — Nova Scotia, New Brunswick, and Prince Edward Island, — whose geological formations are younger, softer, and suited to soil making. Here farms, forests, and mines prevail, and the irregular coast line offers harbors and easy communication with the Old World.

3. Southern Quebec and southern Ontario, lying along the St. Lawrence River and forming the rich peninsula between Lake Huron on the west and Lake Erie and Lake Ontario on the southeast. Here are undisturbed rocks whose wasting makes good soil ; and here also is a cover of glacial drift and lake muds, which offer a permanent basis for agriculture. Here Canada comes down to 42 degrees north latitude, and is like the best parts of New York, Ohio, and Michigan.

4. The plains lying between Lake Winnipeg and the Rocky Mountains, and stretching thence northwest to the Arctic Ocean. The southern part, running northward from Minnesota, North Dakota, and Montana, has in the main enough water for crops, and has vast areas of rich prairie soil. It is believed that agriculture of some types will be highly successful as far north as the Peace River region, as soon as there is good transportation.

5. British Columbia, comprehending most of the Canadian Cordilleras and including the Rocky Mountains in the east, the Selkirks and gold ranges in the center, and the Northern Cascades on the west. The coast is deeply broken, harbors are abundant, the climate is genial through the influence of warm Pacific winds, and the coast and many interior valleys are hospitable to grains, stock, and fruit raising. Fisheries and a great variety of minerals add to the range of commercial possibilities in this province.

235. Forests and forest products. The commodities will be taken somewhat in the order of their development. The forest areas of Canada amount to about 835,000 square miles, or almost one fourth of the country. Before the white man came, the unforested regions were the prairies of the great western plains, the barren lands of the Arctic, and the high mountain

FIG. 164. Douglas firs, British Columbia

areas of the west. There still remain considerable forests in the Maritime Provinces and the Great Lake region, especially in New Brunswick and Nova Scotia. The subarctic forest belt runs from Quebec and Labrador south and west of Hudson Bay, in a north-west direction toward Alaska. On the plains this belt is about 200 miles wide, between the prairies on the south and the barren grounds on the north. The coast regions of British Columbia share the Douglas spruce and other big trees with similar forests extending through the Pacific states of the republic to the south. British Columbia has about one third and Quebec about one fourth of the Canadian forests.

Eastern Canada has large supplies of hard woods. Pine is the most valuable soft wood, and has been largely removed. Other soft woods, especially spruce, find increasing use for pulp and paper in Nova Scotia, New Brunswick, and Quebec; and with proper use this industry may be permanent, as a crop of pulp wood is said to grow in thirty years. Although a young people, interest in forestry is strong, and there is a Canadian Forestry Association which distributes literature, holds conventions, and provides lectures. It is proposed to reforest certain older districts as well as to plant forests on the prairies. A conservation commission has recently been formed by the Dominion Parliament, and there are departments of forestry in the universities of Toronto and New Brunswick.

236. Furs and fisheries. The early history of Canada is almost the same as the story of its fur trade. The discoveries and first settlements in its unending forests were due to zeal in the quest of the fur-bearing animals, — ermine, sable, mink, and beaver. The Hudson Bay Company, founded in 1670, was for generations the real government of the colony, and as late as 1869 sold large rights to the Dominion of Canada. It still exists as a commercial company, with large houses in all the principal cities. Edmonton is the point of supply and departure for hunters, and Winnipeg is a depot for furs.

Fisheries are developed on the Atlantic and Pacific borders and on many inland waters. Since early days the maritime region in the east has been an important fishing ground for England, France, and America. The shallows of the Newfoundland Banks, the Gulf of St. Lawrence, and the borders of Nova Scotia and New Brunswick are the home of cod, mackerel, shad, halibut, haddock, and lobster. Eastport, in Maine, has a large industry in preserving New Brunswick herring, which are

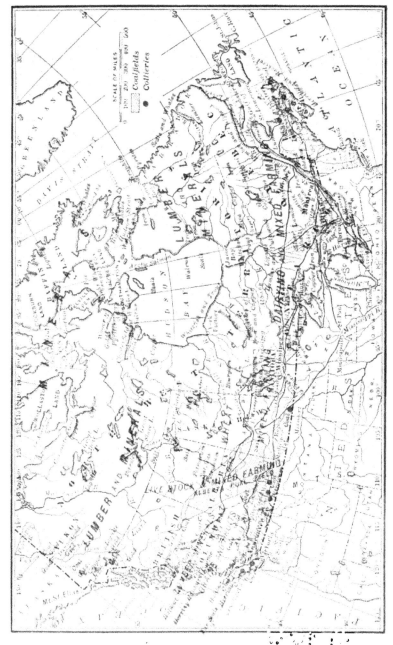

FIG. 165. Resources of Canada

sold as sardines. British Columbia now surpasses even Nova Scotia in its fishing output. It has sea fishing in its fiords and deep bays, with salmon, trout, and char in its inland waters. Best known and most profitable is the salmon industry. The fish are caught as they run far up the Fraser and other rivers, are then canned, and modern transportation gives them a wide market. Seal fishing is also, in particular, an industry of British Columbia and thus contributes another item to the fur trade.

FIG. 166. Salmon catch, Fraser River, British Columbia

The lakes and rivers of Canada on its southern border and throughout its interior offer vast fishing grounds. Manitoba, best known as a wide wheat-growing prairie, has 20,000 square miles of fishing waters, an area as large as Lake Huron, worked by 2000 regularly employed fishermen.

237. Mineral wealth of Canada. The country has vast resources, largely unexplored, and the known regions but partially worked. The mineral product in 1907 amounted to $86,000,-000. In this coal led, with $24,000,000; while of iron, the other basal mineral of commerce, the output (pig iron) was $9,000,000, and this in part was from imported ores. Of native material, iron was exceeded by copper, nickel, silver, and gold.

In the nickel of Sudbury, near Lake Huron in Ontario, and in the asbestos of eastern Quebec, Canada leads all other countries.

Three regions contain most of the coal, — (1) Nova Scotia and Cape Breton Island in the east, (2) Vancouver Island and other parts of British Columbia in the west, (3) widespread deposits of lignite underlying the plains east of the Rocky Mountains. Thus all important sections are near coal, except the settlements and cities along the Great Lakes. Toronto and other cities, even as far west as Winnipeg, import supplies of anthracite coal from Pennsylvania. It comes by rail to the southern ports of Lake Ontario and Lake Erie, and thence by water as near as possible to its destination. Anthracite coal occurs at two Canadian localities, one in southern Alberta and the other on Queen Charlotte Island.

Iron is most worked in Nova Scotia, but there are ores in the Lake Superior region and in British Columbia which are important and will come into use with growing population, better transportation, and the improvement of methods of mining and reduction. Gold is worked in Nova Scotia, in the Lake Superior district, in British Columbia, and in the Yukon region, especially the Klondike district, and thus has a wide distribution in the Dominion. The copper output in 1907 was over $11,000,000, derived chiefly from the Great Lake region and British Columbia. This metal will be useful in the electrical development of Canadian water power, and this in turn may atone for the lack of coal in the Great Lake region, serving in the smelting of iron and for general manufacturing and domestic use.

238. Agriculture. The value of Canadian field crops in 1908 was $432,000,000. The somewhat surprising fact is that oats surpassed in value the wheat, and hay and clover exceeded either of those grains. Barley and potatoes are large products, but far below any of the first three named.

The country may best be taken by regions. The Maritime Provinces have long been cultivated in some favored parts, such as Prince Edward Island, the district around Annapolis in

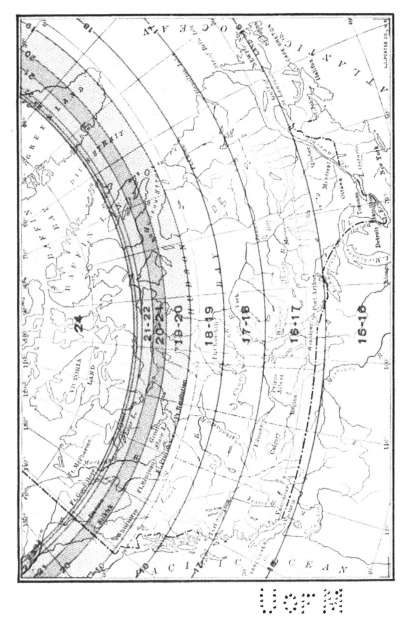

FIG. 167. Latitude belts in Canada, showing the average number of possible hours of sunshine in June

Nova Scotia, and the St. John valley of New Brunswick. Apples and potatoes are typical products in the east. Southern Quebec along the St. Lawrence, and especially all southern Ontario, are the regions of an advanced and diversified agriculture in Canada. Ontario has been an important wheat province, but the increase of wheat in the west has diminished the product in the east, which sees a corresponding growth in live stock and the dairy. The lower lake region of Ontario raises

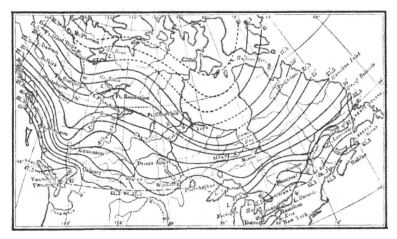

FIG. 168. Canada; isotherms for the year. Broken lines are conjectural

apples, peaches, grapes, and other fruits in abundance and perfection, and this is the one part of Canada which can most successfully grow Indian corn.

Wheat growing is now extending rapidly from the United States boundary across Manitoba, Saskatchewan, and Alberta, and in its beginnings has gone as far north as the Peace River district. Owing to northern latitude daylight is prolonged in the summer, and the climate is softened by the "Chinook" winds, which descend from the Rocky Mountains, the air being warmed by compression as it descends. The effect is seen in the bending northward of the isotherms, as may be observed in fig. 168. At the same time the Canadian Experimental Farms devote much attention to breeding early ripening wheats suited to these provinces, and the crop is already an important factor in the world's markets.

British Columbia is a large province, almost 400,000 square miles in area. Although so largely mountainous, abundant lands in the valleys are suited to tillage. The coastal parts are so moist and warm that all kinds of temperate-latitude fruits are favored, and this type of agriculture will rule. Apples, peaches, pears, as well as grapes and other small fruits, flourish, and at the same time there is room for grain, stock, and poultry, making a typical mixed agriculture. The growth of the coast cities, such as Vancouver, and the new Pacific trade will afford an increasing commerce.

The Canadian government keeps seven experimental farms, the central farm being at Ottawa. Besides the wheat breeding already mentioned, every kind of investigation that could be useful to the farmer is carried on. The extent of this work can be understood when it is stated that in a single recent year nearly 100,000 letters of inquiry were received, and were answered either by letter or by printed information.

239. Manufactures. Manufacturing industry is successful when there is plentiful raw material, available power, good transportation, and a large home and foreign market. Canadian raw materials, drawn from forests, mines, and the soil, are ample. The distribution of coal has been noticed, and of water power it may be said that it is abundant everywhere except on the great prairies of the central lowland. Canada has its share of the power resources of Niagara, while British Columbia and the wilderness between the Great Lakes and Hudson Bay is full of falling water. The province of Quebec alone is estimated to have 17,000,000 horse power.

Much of the manufacturing is relatively simple, as is proper to a new country. Examples are butter and cheese making, flour milling, the canning of fish, and the sawing of lumber. Iron manufacture is in its infancy, but is sure to increase, since machinery, steel rails, and tools are precisely what a new land must have for its development. Cotton, woolen, and leather work, furniture making, and sugar refining have attained importance in the older provinces. The manufactures even of the young city of Winnipeg gave a total of $22,000,000 in 1907. Manufactures are most developed in the older and larger centers of the east, as at Montreal, Toronto, Quebec, Kingston, Hamilton, and St. John.

240. Cities. These may be conveniently divided into four groups, — Atlantic, Pacific, lake, and interior cities.

1. Atlantic cities. Halifax and St. John are by the open sea, while Quebec and Montreal, on the deep estuary of the St. Lawrence, are, for the purposes of commerce, marine cities. By

FIG. 169. Halifax harbor

reason of good harbors or commanding sites all were founded in early days. Halifax has been called the most English city in America. Its population is about 55,000, it has a superior harbor, and until recently was occupied by a British garrison. It is the eastern terminus of the Canadian Pacific and of the Intercolonial Railway, and has long had transatlantic steamship communication. St. John is the chief shipping point for New Brunswick products, and is the winter port of some Atlantic lines. It has coastwise connections with Halifax, Portland, and Boston.

Quebec stands on a noble promontory on the north bank of the St. Lawrence, and is perhaps the most historic city in North

America. The Indian town Stadacona preceded it, and here Quebec was founded in 1608 by Champlain. In 1759 the victory of Wolfe over Montcalm forever transferred Canada from the French to the British flag. Much of its former sea trade went to Montreal when the St. Lawrence was deepened to receive large ships, but it is believed that the increasing size of vessels may bring back some of the city's former importance as a port. It was formerly a center of timber industry, and the development of wood pulp is reviving its importance as a depot

FIG. 170. Montreal harbor

of forest products. The bridge over the St. Lawrence, whose completion was delayed by a disaster, will give Quebec the advantage of being on a direct line of traffic from the west to St. John and Halifax.

Montreal, having 454,000 people, is the largest city in Canada. It is at the head of navigation for large ships, but for smaller vessels its waterways lead up the St. Lawrence to the Great Lakes, and up the Ottawa River to Ottawa. It is a terminal, present or prospective, of all the transcontinental lines of railway, and has easy communication with New York City by way of the Champlain and Hudson valleys. Its immediate hinterland is the most fertile parts of the provinces of Quebec

Canada in population and in commerce. It not only has connection with the upper Lakes by the Welland Canal, but, by a short line of railway, is joined to the ports of Georgian Bay, and thus with the Canadian steamship lines on Lake Huron and Lake Superior. Toronto and Montreal have special distinction among the educational centers of Canada. Hamilton is at the west end of Lake Ontario, is on the line of railway communication with the United States, and its hinterland is the garden ground of eastern Canada. Niagara power is now being utilized, and here, no doubt, as on the American side, an important industrial center will develop. Similarly, on the Canadian as on the Michigan side of the St. Marys River, industry already established will become important, since the place has water power, and transportation both by lake and rail. Port Arthur and Fort William on the north shore of Lake Superior, side by side, are the lake terminals of the Canadian Pacific and Canadian Northern railways, and are towns of swift growth and increasing commercial and industrial importance.

4. Interior cities. Of important inland centers the oldest is Ottawa, the capital of the Dominion. Its growth, like that of other seats of government, is not due especially to its position as a capital city, for geographic conditions give it an industrial and commercial character. It is at the head of navigation on the Ottawa River and is joined by the Rideau Canal to Kingston and Lake Ontario. The Chaudière Falls furnish water power, which is applied especially in the lumber industry. A single plant cuts yearly 110,000,000 feet of lumber, and others from 35,000,000 to 60,000,000 each.

Winnipeg was long the chief post of the Hudson Bay Company, bearing the name of Fort Garry. It is at the junction of the Assiniboine and Red rivers, both navigable streams. More important than this is its position on the border of the prairies between the international boundary and Lake Winnipeg. North of the lake is a country which must remain a wilderness. Every transcontinental Canadian railway therefore must pass through Winnipeg, and all central Canada must be tributary to it. It has about 1 30,000 people, and it has been said that it has "the largest undisputed mercantile territory in the world"; it already rivals, on about equal terms, Minneapolis as a primary wheat market.

Nearly 1000 miles to the northwest is Edmonton, the capital of Alberta, on the Saskatchewan River. It has 30,000 people, and will soon be joined to three transcontinental railways. It

FIG. 173. Winnipeg, looking west along Portage Avenue

has a varied agriculture, plentiful coal, and the Peace River country will be tributary to it. It already has fourteen banks and is the second distributing center in western Canada.

241. Waterways. Canada opens on the east and west to the oceans and continents of the northern hemisphere. Within eastern Canada protected waters bear ocean ships 1000 miles to Montreal. Thence smaller vessels may proceed up the Ottawa to the capital, avoiding rapids by canals. Forty-six miles below Montreal, boats may go by river and canal to Lake Champlain. Up the St. Lawrence, avoiding the rapids by canals, a fourteen-foot passage is gained to Lake Ontario, and the Welland Canal by like depth conducts to Lake Erie. The Sault St. Marie Canal joins Lake Huron to Lake Superior, and the lake terminals for Canada are found at Port Arthur and Fort William.

An important canal is planned to lead from Georgian Bay by an old glacial channel and by present streams to the Ottawa River, giving in the summer months the shortest passage from the upper Lakes to tidewater. The St. Lawrence from Quebec up to Montreal has been deepened to 30 feet, and the work has been sixty years in progress. It is proposed to build a railway to join Manitoba and the west with Hudson Bay, touching its shore either at Fort Churchill or Fort Nelson, thus bringing Winnipeg 800 miles nearer Liverpool than by the present route through Montreal. Navigation would here be open but a part of the year, but it is believed that much of the exported wheat might take this route.

FIG. 174. Snowsheds, Selkirk Mountains, Canadian Pacific Railway

It seems possible in future development to open a waterway from Lake Superior to Winnipeg and Edmonton, using the numerous rivers and lakes, and thence to the Arctic sea by the Athabaska, Peace, Stone, and Mackenzie rivers.

242. Railways. The Canadian Pacific crosses the continent from Halifax to Vancouver, with many spurs and with a network on the prairies. When one sees the long line of Canadian Pacific piers at Antwerp he is at the eastern terminus of a route that leads by ship and rail from Europe to China, more than two thirds the circuit of the globe.

The old Grand Trunk Railway is merged with the new Grand Trunk Pacific, and thus will run from Moncton, New Brunswick, via Winnipeg, 3600 miles to Prince Rupert on the Pacific. The Canadian Northern is a new system begun in 1896, and operating about 5000 miles of road at the present time. It is chiefly in Manitoba, Saskatchewan, and Alberta, where its network occupies a belt north of the Canadian Pacific main line. Its trunk line runs from Edmonton to Port Arthur and a branch joins Duluth to the system. It has lines in central Ontario and will ultimately join the two oceans. Its eastern terminal will be at Quebec. There are nearly 1500 grain elevators in western Canada, and the capacity of the Port Arthur and Fort William elevators is about 15,000,000 bushels. Before the days of the Canadian Pacific road the freight on a bushel of wheat from Winnipeg to Liverpool was six shillings. It is now ninepence and the distance is 4500 miles. A vast population can live across an ocean and across half a continent from the fields where their bread is grown.

243. Foreign commerce. The total foreign trade of Canada in 1908 amounted to $650,000,000. Of this sum $280,000,000 represented exports and $370,000,000 imports. By far the greater part of this trade is with the United States and Great Britain. Canada sends to Great Britain a great supply of foodstuffs; hence these exports are larger than to the United States. On the other hand, the United States sends more than twice as much merchandise into Canada as the mother country. This is done, although a preferential tariff admits goods from Great Britain on more favorable terms than from the United States. The reason is that a kindred people, with similar problems and needs, is our close neighbor along an extended boundary. Iron and steel, coal, woolen and cotton goods, sugar, drugs, and chemicals are the larger items in Canadian imports, while the chief exports are cheese (nearly $23,000, 000 in 1908), cattle, bacon, wheat, wood and wood pulp, silver, gold, and copper. Montreal is the leading city both for exports and imports. Toronto has a large import trade but almost no export business.

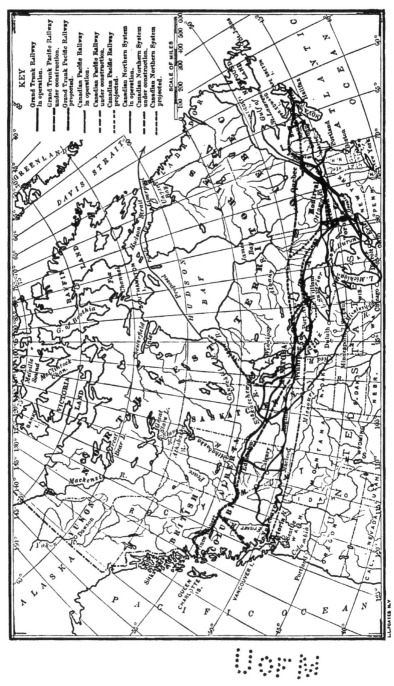

FIG. 175. Transcontinental railway systems of Canada

CHAPTER XIX

244. Basis of commercial greatness. The name "Great Britain" is correctly used of England, Wales, and Scotland, but the designation "United Kingdom" is better, if Ireland be included. By usage, however, the adjective "British" may be correctly applied to the United Kingdom in general. "England" and "English" are often incorrectly used concerning the United Kingdom.

The islands are substantially included between the parallels of 50 and 60 degrees north latitude, and the mild climate is to be contrasted with the stern conditions of the same latitudes in the Labrador and Hudson Bay regions west of the Atlantic. The area of the United Kingdom is 121,000 square miles. The insular position is important for defense and commercial progress. The United Kingdom is so close to continental Europe that commercial exchange is easy, and a short ocean track joins it to the United States and Canada.

There are other reasons for commercial growth, as is shown by Chisholm. The areas of good soil are relatively large and the climate is favorable for labor and productiveness. Neither the heat of summer nor the cold of winter is such as to check endeavor. The rainfall is ample, being above 60 inches in the highlands of the west, and from 25 to 30 inches on the eastern lowlands. Coal and iron are the most important raw materials produced. Good harbors are many, and Britons have for centuries been keen sailors. In the interior, England has physical unity, and close communication binds together all parts of the kingdom. Bays and estuaries run far inland and their heads connect by short routes with each other and with interior cities. Colonies lie in every part of the world.

British character and hereditary qualities, as well as British geography, explain the march of industry and trade. It is these which make labor effective and invention strong. This inner quality early developed thrift and progress, accumulated capital, built up markets and held them, perfected shipping, and diffused the English language. British character took possession of North America against odds, and in regions whose sovereignty

FIG. 176. Eddystone Lighthouse, English Channel

it lost it yet finds the largest market for its manufactures and the greatest source of its raw materials. Geographic conditions but partly explain how a country less than half as large as Texas supports 44,000,000 people, and has, of all nations, the largest shipping, the largest total of foreign commerce, the greatest colonial system, and offers the largest market for the world's merchandise.

245. Metropolitan England. It is useful in this study to follow Mackinder and make a distinction between "metropolitan" and "industrial" England. The former is the southeastern or lowland part, whose center is London. It is the side of England which faces the continent, having little coal and little water power. Outside of London it has no cities of the first order, and but few containing as many as 100,000 people. A line drawn from the Severn to the Wash is substantially its northwestern boundary.

Its present life is of more ancient origin than the life of industrial England. Wealth, conservatism, and social development are here, and these culminate in London, the seat of government, the financial and literary center of the Empire.

246. Industrial England. From the Scottish border, midway between seas, an upland runs south and descends to the plains in the center of England. This broad, low mountain ridge, nowhere rising to 3000 feet and usually of much less height, is the Pennine Range, often called the "backbone" of England. Around this range — on its east side, about its south end, and on its west side — lies industrial England. It is the region of coal and iron. On the east are Newcastle and the county of York. Here is a group of industrial cities, — Leeds, Bradford, Halifax, Huddersfield, and Sheffield, in the basin of the Humber and back of the ports, Hull and Grimsby. South of the Pennine Range are Derby, Stafford, Dudley, Wolverhampton, and, above all, Birmingham with its smoking chimneys and endless rows of the brick cottages of its laborers. On the west is the county of Lancashire, with Manchester and its cluster of satellite towns humming with spindles and looking toward Liverpool and the Atlantic.

In industrial England life is new, and there is more toil and less leisure than in southern England. On the inventions of the past century and a half, and on the coal and iron development of the last century, industrial England is founded. It lives in the atmosphere of trade, its feeling is democratic, and its universities are the outgrowth of present times. It is perhaps typical of the life of the two regions that metropolitan England has one center and industrial England has many.

247. English agriculture. For the American student it is less important to know the details of farm products than to see the relation between agriculture and industry. To 1850 British food, so far as the nature of climate and soil permitted, was raised at home. Now the greater share of it is imported. In 1850 there were in England and Wales less than 18,000,000 people. In 1908 the population was more than 35,000,000. The population has thus almost doubled, and in many agricultural products the movement is backward. With modern methods of tillage and transportation new countries can lay down wheat and meat in English cities more cheaply than the

English farmer can raise them. The land is more and more occupied by roads and houses, and labor is better rewarded in the factory than on the farm. Iron and coal, with skill and industry added, enable the Englishman to support himself by manufacturing and commerce, selling his goods to those who raise his food, and who raise also most of the raw material for his factories.

Oats, barley, and wheat are the largest grain crops. The last is most raised on the eastern lowland, and shows larger average yield per acre than the wheat of most countries. Potatoes and turnips are large crops, and market gardening increases for the supply of the city populations. There is much pasture and meadow, and for a small area the animal industry is large. Dairy interests are important, but the chief supply is from foreign lands, — Denmark, Canada, and Australasia. Less than 4 per cent of the United Kingdom is in forest, and the woodlands serve less for timber than as pleasure grounds for the rich.

248. English cotton industry. The cotton working of the United Kingdom centers in Manchester and the surrounding towns in Lancashire, on the west slope of the Pennines. The moist, warm winds from the sea maintain a high degree of moisture in the air and favor the handling of the cotton fiber. The power for machinery is supplied from the Lancashire coal field. The raw material from the United States, Egypt, and other lands comes in by the neighboring port of Liverpool, and in smaller part direct by the Manchester Ship Canal. By the same means the product of the mills is sent to all parts of the world in British ships. The country supplies everything but the raw cotton.

Manchester and the adjoining Salford have nearly 1,000,000 people. This center is called by Mill mercantile rather than manufacturing, being full of warehouses and being the financial center, while the mills and spindles are in neighboring towns. Of these Oldham has nearly 150,000 people and is said to hold one third of the cotton spindles of England. Bolton, Blackburn, Burnley, and Preston are important centers for cotton goods. At Preston, Arkwright set up spinning frames in 1768. Outside of this district Nottingham is the chief town devoted to cotton, specializing in lace and cotton hosiery. It has 260,-000 people, and, like the towns above named, directs the American student's attention to the size and importance of foreign towns of which he often knows no more than the name.

FIG. 177. Card room showing drawing and roving frames; cotton mill in the Lancashire region

Brooks and Doxey, Manchester

The sources of England's supremacy in cotton manufacture were noticed in sect. 28. Her advantage of an early beginning will be smaller as time passes. Other nations are gaining skill and experience in this industry, and the amount of cotton milled in the United States, France, Germany, and Italy shows that these nations will increasingly supply their own markets. Much will depend on the energy with which the cotton-working peoples seize and hold the market in the populous but backward countries. No doubt the leadership of Lancashire will become less conspicuous, but whether it will pass to some other part of the world depends on many conditions and cannot now be told.

249. English woolen industry. About forty miles northeast of Manchester, on the other side of the Pennine upland, is Leeds. About Leeds are Bradford, Halifax, Huddersfield, and other large towns, and here in Yorkshire is the chief center of the world's woolen manufacture. The population of Leeds was estimated in 1908 at 477,000. Bradford had nearly 300,000

FIG. 178. Picker room, Lancashire cotton mill
Brooks and Doxey, Manchester

and is but eight miles from Leeds. Halifax, with 111,000, is more than twice as large as the Halifax with which the American student is familiar. On the neighboring Pennine pastures sheep have been raised for centuries, and in the villages among the hills handwork in wool was the forerunner of the factories of to-day. The supremacy of Yorkshire in woolen goods is due to a number of causes, among which are the early keeping of sheep in the region, the persecution of skilled weavers on the continent and their immigration to England, the prior invention of machinery and adoption of the factory system, and the concentration of capital and skill which always favors a locality when its industry has been long established and is known throughout the world. According to Chisholm, Leeds is first in the wholesale clothing trade in Great Britain, probably first in leather, and is important in iron and steel. It is natural that such centers as Manchester and Leeds should meet their own demands for textile and other machinery. Many towns in this district engage in special forms of woolen manufacture, and Bradford works in silk, velvet, and plush goods.

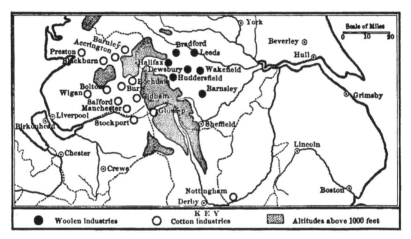

FIG. 179. English centers of cotton and woolen manufacture

Leicester lies some distance to the southward, in the center of England. Woolen hosiery is more largely made here than elsewhere. The population in 1908 was 240,000.

250. English mineral industry (sects. 56 and 66). Excepting iron, the mining of metals is a small industry in England. Cornwall and Devon are the chief European source of tin, and the output in 1907 was 4407 tons of metal, worth nearly $4,000,000. In the same year the exports of tin from the Straits Settlements in British India were valued at more than $82,000,000. Some lead and zinc are mined in the north of England, the product for the United Kingdom in 1907 being worth a little over $3,000,000. The production of copper is not important, and even smaller is the output of gold and silver. Of the nonmetals, building stones and pottery and other clays are abundant. Staffordshire, Worcester, and Derby are centers of potteries, and ornamental stoneware is made in London. Glass is made in some of the coal fields, and salt deposits are well distributed. The total mineral product of the United Kingdom in 1907 was valued at $676,000,000. Of this England produced $448,000,000. Of the total for the United Kingdom coal is credited with about 89 per cent.

251. British fisheries. Reference is here made to this industry for the United Kingdom as a whole. Irish production is small, and British waters are sought mainly by English and Scotch fishermen, this being due to the greater markets of Great Britain and the superior richness of North Sea waters. These waters, like those of the Newfoundland Banks, are shallow. Yarmouth and Lowestoft are among the old fishing towns of the east coast, but the industry is concentrating in larger ports, as Grimsby and Aberdeen. The total product of the fisheries of the United Kingdom in 1908 was valued at $50,000,-000, and about 107,000 persons were employed. With short distances and swift trains fresh fish make an important part of the food in Great Britain, and express passenger trains have been sidetracked to allow fish trains to move toward London, where, at Billingsgate, is the largest fish market in any land.

252. Wales. This region closely adjoins England and is near to such industrial and commercial centers as London, Birmingham, and Liverpool. Here is a land of ancient rocks, much deformed, still mountainous, and often rugged. Agriculture is limited and of purely local interest, and coal far outweighs in value all other products, being found in the South Wales coal field, adjacent to Bristol Channel and extending northward from Swansea and Cardiff. The population of Cardiff in 1908 was about 190,000, and its industries were in iron, steel, tin plate, and shipbuilding. Swansea has about 100,000 people, and its industries are in tin plate and the smelting of copper. Wales has some iron, but the ores are mainly imported from Spain, and the copper ores are brought to Swansea from many countries. Roofing slates are much quarried in North Wales.

253. English and Welsh seaports. The seaports will be taken in groups, and first those of the east. coast. London is the first of British ports. It is the center of the British Empire, the focus of the railroad systems of Great Britain, and opposite the mouths of the Rhine and Scheldt, important navigable rivers of the continent. Not being in the industrial district, its imports are greater than its exports. It is a distributing center. First, its own population requires large supplies. Second, it distributes foreign commodities to all parts of Great Britain by the railways and by the coasting trade, having more of the latter than any other port. . Third, it reëxports merchandise from all parts of the world to continental countries, but this business is now somewhat narrowed by the growth of foreign trade at Bremen, Hamburg, and other ports of the continent. Like most great British ports, London is inland, being on the estuary of the Thames, at the head of navigation.

Fig. 180. Looking south across the landing stage and the Mersey, Liverpool

The next deep reëntrant going north is the Wash, but this is commercially unimportant. Then comes the estuary of the Humber, leading to Grimsby and Hull. The latter is the third port of the United Kingdom in the importance of its shipping. It is an outlet for the woolen and iron of the Leeds and Sheffield regions, and reaches across the Pennines and carries out some of the cotton of Lancashire. Its foreign connections are with continental nations, India, Australia, and the United States. Newcastle upon Tyne is the great port of the north of England. It is eight miles from the sea, from which the docks are continuous, and the trade is in coal and iron, shipbuilding being a large industry.

Of ports on the west coast Liverpool is the first. It is on the estuary of the Mersey, which receives the largest ships. The large city of Birkenhead, south of the Mersey, belongs to the same commercial center. Liverpool is the great port for the United States, and has a vast import trade in foodstuffs, cotton, iron, lumber, and tobacco. Its lines connect with Africa, the Mediterranean, Australia, and India. Liverpool and Birkenhead had in 1908 about 900,000 people.

Bristol is an ancient port, and is said to have been, in the fourteenth century, second only to London. It has declined in recent times, but new docks have been built in an effort for the revival of its foreign trade. Cardiff and Swansea have already been named as places of industry. Their importance in foreign trade lies especially in exports of coal, which are very large, and imports of ores of iron and copper. Holyhead on the Welsh coast is significant as a point of transfer for London mails to America, via Queenstown and the Atlantic mail ships. Recently also Fishguard has been made a port of call for some of the fast ships of the Cunard Line.

Southampton is the most important port on the south shore. Several Atlantic lines make this port for London, and some White Star Line ships now come to Southampton in place of Liverpool. Plymouth, in Devonshire, has considerable foreign trade, and has long been one of the chief British naval stations. Transfer to the continent is made from New Haven, Folkestone, and Dover on the English Channel, and Harwich on the North Sea. Dover lines run to Ostend in Belgium and Calais in France.

254. English railroads and canals. Where the Romans planted civilization in England nearly 2000 years ago, they used many of the routes of travel over which modern roads now pass. Such routes ran out from London, and to-day in a marked way the railways radiate from the metropolis. On the south and east they are short because the sea is nearer, but on the west and north they run to Lands End, to Bristol and the Welsh ports, to Liverpool, the great industrial towns, and the Scottish border. The roads are well built, there are no grade crossings, on many through lines great speed is made, and owing to the dense population and many large cities passenger travel is enormous, but the number of accidents, by reason of careful maintenance and management, is small. The railway mileage of England

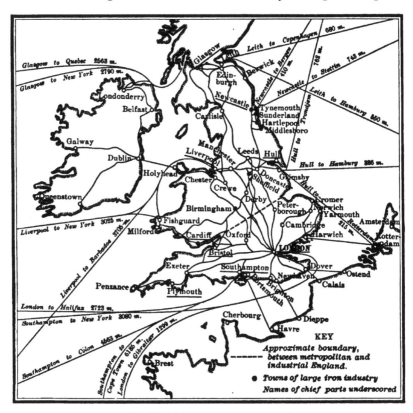

FIG. 181. Chief British railways, ports, and sea routes

and Wales was 15,897 at the beginning of 1908. The total for the United Kingdom was 23,108. The freight or "goods" traffic of England differs from that of the United States in the small cars used. These better suit the short-distance carriage, the local trade, and vast number of small consignments.

England and Wales have 3641 miles of canals, of which 1184 miles are either owned or controlled by the railways.

England had its era of canal building a hundred years ago, and, as in America, such waterways have declined, the railways are charged with destroying their usefulness, and present agitation looks toward their revival. Most of the commercial rivers are joined by canals crossing the central plains of England.

The Manchester Ship Canal is an important modern construction opened in 1894. Its length is 35½ miles and its depth 28 feet. It connects both with the sea and with the large canals of the interior, and is designed to release the cotton and other trade of Manchester from subjection to the shipping interests of Liverpool. Both post and telegraph are under government control.

255. Scotland and Scottish agriculture. Scotland has a central lowland and two highlands. The lowland stretches from the North Sea to the Atlantic, and is deeply indented by the Firth of Forth and the Clyde. The greater highland is on the north, is cut by deep glens, and has on its northern and eastern fringe some areas of fertile lowland. Like the lowland, it is deeply indented by fiords and sea lochs, and is bordered, especially on the west, by innumerable islands. The higher grounds in the south may be called the Scottish *upland*, leaving *Highlands* as a distinctive term for the north. This region is less high and rugged than the Highlands and is largely used for sheep pastures. Railways from England follow the east and west coasts and also thread the center of this upland.

The higher grounds and islands of Scotland make 70 per cent of its territory and contain but 23 per cent of the inhabitants. Most of the agriculture as well as the industry is in the thickly populated central lowland extending into the interior from Edinburgh and Glasgow. Three fourths of the grain crops are in oats, barley being the remaining cereal of importance. Here is seen the influence of a northern latitude with damp and cool climate. The days of summer sunshine are, however, long. Most of the plowland is in the central lowland. Potatoes, turnips, and swedes are grown, and roots are said to be as important for stock as corn is in the United States.

In the mountainous districts the population is scattered and the grazing of sheep and cattle is the chief interest.

256. Scottish industries. As in England, textiles have a large place, and this applies to cotton, woolen, and linen goods. The Scottish Lowlands have in close association coal, iron, and shipping facilities, with intelligent and industrious labor and inventive skill, a description which answers precisely for industrial England. The cottons center about Glasgow. The words "tweed," "tartan," and "plaid" suggest types of Scottish

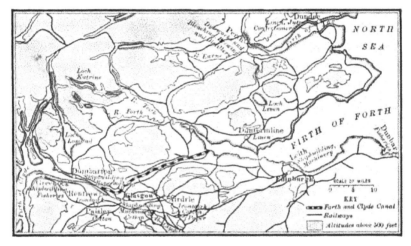

FIG. 182. Industrial and commercial map of the Scottish Lowlands

woolen manufacture which are everywhere known. The distilling of whisky is a large industry, and a third class of manufactures relates to iron and steel. To the last belongs the shipbuilding of the Clyde, which is unequaled. Another industry of the Lowlands is made possible by the occurrence of shales containing mineral oil, the extraction of which affords a considerable product.

257. Scottish seaports. The greatest of these is Glasgow, which in 1908 contained about 860,000 people, and is the second city in the United Kingdom. The development of adjacent coal and iron and the deepening of the Clyde have gone on together during the past century of new use of fuel, machinery, and ocean trade. As with Liverpool, the trade of Glasgow has grown largely out of commercial relations with America, and many steamships from the St. Lawrence make port there. Looking at a general map of Scotland, Glasgow seems to be in the heart of the country, so far does the Clyde penetrate the interior. This in less degree is true of Edinburgh and Leith, and thus the two centers of Scotland are in close relations both with the sea and with the Scottish lowlands.

Leith is on the south shore of the Firth of Forth, two miles from Edinburgh, of which it is the port. The combined population is nearly 450,000. Edinburgh is not an industrial city, but the importance of the port is considerable, and it has shipping relations with Germany, Denmark, the Netherlands, and other countries. Considerable exports of coal and iron here reach the sea. Dundee also has an estuarine harbor along the Tay, fifty miles from Edinburgh to the northeast. It is third in importance, of Scottish ports, having industrial work in linens, jute, confectionery, and jams, and is the center of considerable foreign trade. Its population in 1908 was 168,000. Aberdeen, on the east coast, north of Dundee, has its chief industries in cotton, iron, paper, and the quarrying of the local granite. Of the last the city is built, and it forms an important part of its outgoing freight. Thus Leith, Dundee, and Aberdeen, though smaller, compare with London, Hull, and Newcastle on the east coast of England, and Glasgow dominates the west coast, as does Liverpool in England.

258. Ireland. This part of the United Kingdom holds small place in the world's commerce, since its trade is mainly with Great Britain. This appears in the fact that its two chief seaports, Dublin and Belfast, are on the east coast. Ireland has had its largest effect on commerce and industry by the multitude of emigrants sent to other lands. This is especially true of the United States, to which Irishmen formerly came in great numbers as laborers. Their descendants are now widely engaged in trade, manufacturing, and agriculture, and form an important element in the American population.

Ireland consists of an interior, well-watered lowland, bordered about the sea with low mountains. Some rivers penetrate the lowland to its center and several are joined to each other by canals, giving a complete passage of the island, as in England. The river Shannon flows from the north, southward through the heart of Ireland to the Atlantic on the west. Its upper waters are canalized, and there is canal and river connection with Belfast in the northeast, Dublin in the east, and Waterford in the southeast.

Agriculture rules in Ireland, with potatoes, oats, and roots for home consumption and stock raising; cattle, horses, and poultry being exported to Great Britain, as also eggs and dairy products. Flax is grown in the north in connection with the linen industry. Manufactures are limited, owing partly to ancient social and governmental conditions, and partly to small supplies of iron and coal.

259. Commerce of the United Kingdom. The United Kingdom is essentially a free-trade country. Customs duties are collected only upon the following articles : tea, coffee, spirits, wine, tobacco, sugar, dried fruits, cocoa, and chicory. In 1907 the imports on which duty was paid amounted to about one twelfth of the total imports. The raw materials of manufacture, manufactured products, and foods and food materials nearly all enter free. The total imports and exports in 1908 amounted to £1,050,000,000 sterling, or over $5,000,000,000. The import trade is considerably greater than the export trade and is different in character. The following classes of merchandise were imported to values above £10,000,000 each ; grain and flour led with £72,000,000 ; then followed cotton, wool, sheep and lambs, meat, sugar, butter, wood and timber, silk manufactures, coir, flax, hemp, jute, tea, chemicals, dyestuffs, fruits, and hops.

The exports tell a different story. Far in the lead are textiles, including cotton manufactures and cotton yarn, with a value of £94,000,000 sterling. Woolen and worsted come next with £28,000,000, and linen and other materials make

£17,000,000. Metal work, ships, machinery, and hardware contribute £77,000,000 in value, and coal adds £41,000,000, being so far the world's greatest export of coal, and much the largest export among native products of the United Kingdom.

Many facts concerning the distribution of British commerce among the nations have already been given in the chapters on wheat, cotton, animal industry, iron, coal, and in the chapters on Canada and the foreign trade of the United States. It may now be stated that the United Kingdom, with its mercantile marine surpassing that of all nations, with colonies in every part of the world, with the need of importing most of its food and raw material, and with high industrial persistence and skill, has built up the most widely distributed and largest trade of any nation. Whether this position can be maintained, can only be revealed by the future. It is significant that in recent years the United States has exceeded the United Kingdom in the total value of manufactures and in the total of exports. This is not surprising in view of the size of the republic and the vast resources found within home territory.

Some causes of British supremacy in industry and commerce may here be summarized. The discovery of America placed Great Britain at a pivotal point between the Old and New Worlds. British character and skill in handicraft became effective during the modern centuries. Britain at the critical period was the pioneer in the invention and use of machinery, and was the first to develop coal and iron. Transportation was favored by her insular position, her harbors, tidal rivers, and early development of canals and railroads. Her mastery of the ocean gave her raw materials from all countries and made the world her market. Her adoption of the free-trade system resulted in cheap food and raw materials, and enabled her long to undersell competitors in the great lines of manufacture. The extension of the Empire made her shipping safe, gave outlet to her surplus people and capital, and added security to her markets.

Britain's competitors, the United States and Germany, now share with her many of these advantages and at present surpass

are
o,
he
m.
ce
on
rs
av
le
n
od
l,
r
e
e
l

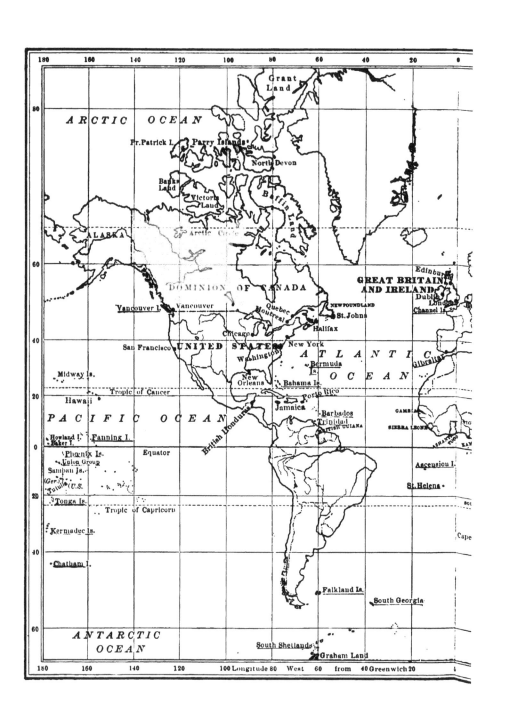

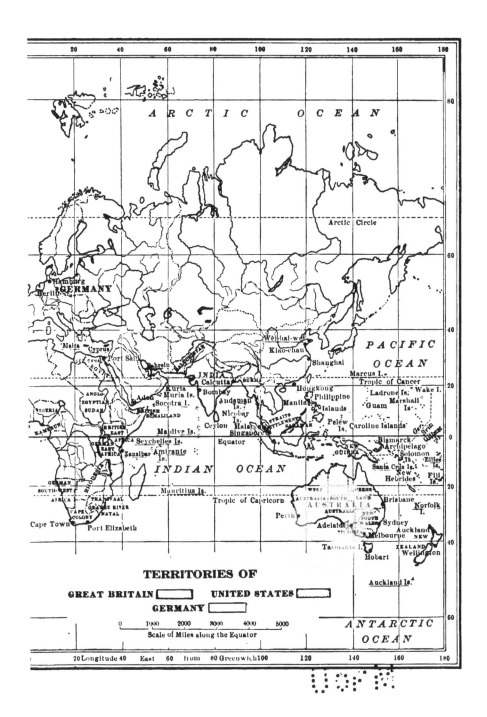

her in rate of growth. Whether one nation or another is supreme is perhaps more important to national pride than to the world's progress. The high rate of British taxation required to meet the interest of her debt, the support of her navy, and for other public ends, may handicap progress. There is need also of a more thorough and general technical education, in which both German and the United States are at present more advanced. The future also will decide how far the Empire can have real unity, and to what extent great and remote states, like Australia and Canada, will hold to the mother country not only in sentiment but in trade. Great Britain being a small country with a large population must buy the world's raw material and sell in the world's markets. So long as her stores of iron and coal last she may continue to be a great industrial nation, whether supreme or as one among others.

260. British Empire. The chief British dependencies are here noticed because they give to the United Kingdom large commercial opportunities. Some have a temperate climate and are largely settled by Britons ; such are Canada, Australia, New Zealand, and, to some degree, South Africa. These colonies are peopled and developed by men born in the British Isles or of British ancestry. They give the mother country favorable terms of trade and supply her with food and various raw materials in exchange for the manufactures which they need. Such trade stimulates navigation and meets all the great needs of the homeland. Keltie refers to colonies of *settlement*, like the above, and colonies of *exploitation*. The latter are usually tropical, are unfitted for the permanent homes of Europeans, and are therefore only governed by the British, while the work in field and factory is done by natives. Such is India, a region of vast population and large resources, giving the United Kingdom wheat, cotton, and other things, and buying her manufactures. In the same class is Ceylon, sending tea, coffee, coconuts, spices, and graphite. Here belong the United Kingdom's colonies in the East Indian Islands ; also the island of Mauritius and colonies and dependencies in central and northern Africa, including

Egypt. We must add also British possessions in and about the Caribbean Sea, British Honduras, British Guiana, with Jamaica and other islands.

A third class of colonies is of commercial importance. Here belong small islands and other posts of defense and supplies. A vast merchant marine and navy need coaling stations in every part of the world. Such are Gibraltar and Malta in the Mediterranean ; Bermuda, St. Helena, and the Falkland Islands in the Atlantic ; the Fiji and many other islands in the Pacific.

These outposts are strongly fortified and give to the navy of Great Britain convenient bases of supply in all parts of the world. They are also in communication with London by telegraph lines and ocean cables, and thus promote the unity and coöperation of all the countries of the empire.

CHAPTER XX

261. The French land. France has an Atlantic shore and ports (Bordeaux, Nantes) facing America; it borders the English Channel and the North Sea, and thus reaches Great Britain and all other countries of northern and central Europe; it also has a shore line and its largest port (Marseilles) on the Mediterranean, and thus has a way to every Mediterranean country of Europe and Africa, and, by the Suez route, to the Orient. Concerning the land borders of France one of her geographers has said that the country is "encircled but not imprisoned." The high and unbroken Pyrenees set a wall on the side of Spain, but roads follow the sea at the ends of the range. The Alps stand on the borders of Italy and Switzerland, and the western Alps, with Mont Blanc, are in France. Railways follow the sea along the Riviera and pierce the Alps by the Mont Cenis Tunnel, thus leading to Genoa, Turin, and Rome.

Lower mountains lie on the east and north. The Jura is between France and northern Switzerland, the Vosges is on the border toward the middle Rhine country of Germany, and the Ardennes is the bounding upland between France and Belgium. The ways of commerce lie through gaps between the mountains. A railway passes from Paris and Lyons up the Rhone to Geneva and the heart of Switzerland. Railways and a canal go north of the Jura and south of the Vosges, by Belfort to Germany and the Rhine. Other gaps lead across the northeastern Ardennes and the Rhine highlands.

The central plateau is more in the southeast than in the real center of the country. It rises boldly in the Cévennes, west of the Rhone, culminating in Mont Lozère at an altitude of between 5000 and 6000 feet. It slopes westward and northward and interposes an area of poorer soil and thinner population in the heart of France. It is the divide between the Atlantic and Mediterranean rivers of France. The Saône and Rhone flow at its base on the east, and the Garonne and Loire flow down its western slope. The Seine is the river of northern France, while the Moselle and the Meuse are important streams finding their lower courses in Germany and Belgium.

The latitude ranges from 42.5 degrees to 51 degrees, Paris standing at about 49 degrees, — as far north as the boundary between the United States and western Canada. Marseilles has about the latitude of Boston, and the contrast in their temperatures shows the varied control of climate. The presence of seas moderates the climate and gives an average rainfall of about 30 inches.

262. Agriculture. Mild climate and good soils give France a high position in the tillage of the ground, the chief interruption being in the central plateau and the mountain borders. About one half of the country is cultivated, and of this, one half, or one fourth of the whole, is given to cereals. Halving again, one eighth of the land is devoted to wheat. In sect. 10 it was shown that in the years 1901–1905 France raised 97 per cent of the wheat used. In 1907 the crop was 368,000,000 bushels, which supplied home needs and left 25,000,000 to 30,000,000 bushels for export. This yield is about half that of the best years in the entire United States, and shows how highly each French farmer cultivates his few acres. The invasion of low-priced wheat from new countries is prevented by a prohibitive duty of 37 cents per bushel on wheat and $2.75 per barrel on flour. Thus the duty is much higher on flour, which, when importing is necessary, favors the French miller. France contrasts with the United Kingdom, which admits wheat and flour freely, raises little wheat at home, and buys it with the products of the mill, the factory, and the shipping. The farms of France are commonly small and worked by the owners. This leads to more effective cultivation than in lands where holdings are large and cultivation is by tenants. It is nevertheless true that on the selected lands devoted to wheat in England the yield per acre is much larger than in France. Agricultural education is carefully promoted in France under government direction.

Fruits next to cereals are typical products of French soil, and among these the grape is chief, flourishing almost everywhere except in the north. Even to this there is an interesting excep-

tion, for champagne wines are produced at about 49 degrees north latitude; and the vice consul at Rheims, the center of this trade, reported, for 1907, exports of this wine to the United States valued at $5,357,-ooo. Champagne is the name of an old province, now making several departments. From this district one may go south by the Saône-Rhone valley to the Mediterranean and

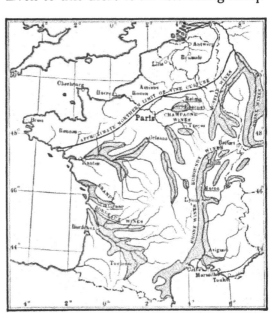

FIG. 183. Chief wine-producing areas of France

Names of wine and brandy centers underscored

be ever in the domain of the vine. Going around the south end of the central plateau, he may then follow the Garonne past Toulouse and Bordeaux to the sea and he will find vineyards abounding. Burgundy, Médoc, and Bordeaux are French geographic names better known through wines than for any other reason.

Olives are also raised in the south, and many nuts, chestnuts, and walnuts, the latter being an important item of the export trade to the United States. Mulberry trees are grown, the leaves serving as food for silkworms, and the growing of sugar beets is in an advanced state.

FIG. 184. Lyons, showing junction of the Rhone and the Saône

324

263. Minerals. Coal and iron, the two substances basal to industry, form larger interests than in any other European countries except the United Kingdom and the German Empire. There are many deposits of iron, but the most important are in the extreme north, along the Moselle adjacent to Belgium. The principal centers for making iron and steel are Lille in the north and Le Creuzot in east-central France. Of the coal 60 per cent is likewise mined in two northern departments adjacent to the English Channel and the Belgian border. Another coal field is in the department of Loire, in the central plateau, west of Lyons and tributary to the industrial center, Saint-Étienne. The coal fields of France occupy 2100 square miles. A considerable amount of coal is imported each year. More than 10,000,000 tons of British coal were brought to France in 1907. Building stones and cements are abundant, and the clays suggest the well-known centers for fine pottery, — Sèvres near Paris and Limoges in west-central France.

264. Fisheries. More than 150,000 people are engaged in this industry, and nearly every port on the Channel, on the Atlantic, and on the Mediterranean has a fishing trade. French boats ply on the Channel and the North Sea, and go far across the Atlantic to Iceland waters and to the Banks of Newfoundland, where Frenchmen have fished since the time of the early voyages of discovery and the first planting of settlements in the New World. In these two distant fisheries 142 French vessels were engaged in 1907. Their cargoes were chiefly discharged at Bordeaux and amounted to 55,000,000 pounds. Oyster growing is an increasing industry along the Atlantic coast.

265. Manufactures. Like the United Kingdom, France imports considerable iron ore, but, unlike her northern neighbor, she also imports much coal. Iron manufacture, therefore, while advanced, is not of the first order. Wine offers a manufacture simple in kind but large in quantity. The same may be said of flour making, which is a great industry, since so much wheat is raised and consumed ; and if any is imported, the greater duty on flour insures the importation in the form of grain.

The textile and clothing industries reach a high total, the textiles alone making about $600,000,000 per year. The woolen, cotton, and linen industries are found more largely in the northern

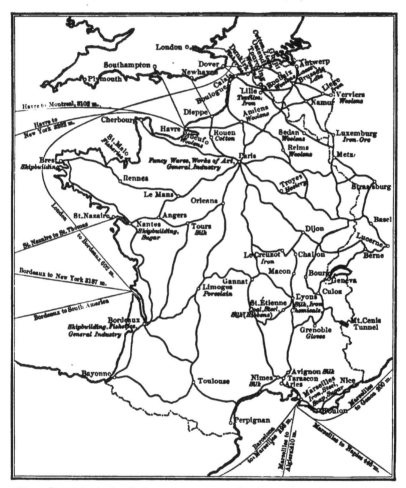

FIG. 185. Principal railways, sea routes, and industrial centers of France

departments. As much of the coal and iron is here, we have a field of general industry comparable to northern England, while in the south, as in England, agriculture prevails. The district in and about Lyons, however, offers an exception, since it lies toward the southeast. Lyons is the third city in France in population and is the first of all centers in the silk industries of the world. It is at the eastern base of the central plateau, at the junction of the Saône from the north and the Rhone from the east. The Rhone-Saône valley has, since ancient times, been a highway from the Mediterranean to northern Europe; hence Lyons occupies a natural position for a city. Much raw silk is produced in the Rhone valley, but more is brought from Italy and the Orient. Lyons has been an important industrial city since Roman times, and in addition to silk has trade in grain and wine. It is a railroad and financial center, and is the focus of much interior navigation. The making of garments, ornaments, and notions is the characteristic part of French industry, drawing upon the taste and hereditary skill of the people. For such products Paris is preëminent. An interesting change seems to be taking place by the use of water power for electrical development in southeastern France. In a single recent year 150,000 horse power was thus distributed, effecting a needed saving in coal and making it possible to divide the work even among homes instead of massing it in factories. It is said that 8000 such purchasers of power are at work in the manufacturing center of Saint-Étienne in the uplands west of Lyons.

266. Transportation. With one exception the chief railroads of France radiate from Paris much as those of England center in London. The most important road is known as the Paris-Lyons-Mediterranean line, and as it reaches Marseilles it joins the three great cities of France, threading the heart of the country, receiving the produce of its greatest industrial centers and many of its richest fields. From it branch lines to Geneva and central Switzerland, also to all parts of Italy.

From Paris other lines reach Toulouse in the south, Bordeaux and Nantes on the Atlantic, Havre and all other Channel ports, and the industrial centers of northern France, while to the east lines pass across the German frontier.

All the chief rivers have been improved, and they afford 5300 miles of navigation. There are also 3000 miles of canals, and these are so placed as to join the headwaters of the chief rivers and offer a complete network for crossing the country from east to west and from north to south. Thus a canal boat could go from Lyons (*a*) down the Rhone to the sea ; (*b*) to Strassburg and the Rhine ; (*c*) to Antwerp or Rotterdam ; (*d*) to Paris and

FIG. 186. Interior waterways of France

Rouen ; (*e*) to Tours, Nantes, and the Atlantic. One fourth of the internal trade is by waterways. Tributary to railways, waterways, and cities, the public highways of France have been highly developed and are kept in an exceptionally perfect condition.

267. Paris. The capital city is in northern France, near the sea and the industrial districts. At the same time it communicates readily with the south and east, and is on the usual route between French and English ports of the north and west, and the interior and southern countries of Europe. It is on the Seine, below the junction of the Marne, and is surrounded by the rich country known to geologists as the Paris Basin. Its local hinterland is one of the richest in France, and the population of the city in 1906 was 2,763,000, making it the third of the world's great cities. It is the first of the industrial and commercial cities of France and is characterized not by one industry but by many. Thus as a center of general industry it is like New York and London. Fashionable clothing and articles of luxury belong to the industry and trade in a special manner. In 1907 Paris sent to the United States merchandise to the value of $63,000,000, or almost half of all French exports to this country. Among the chief items were the following : works of art, costumes, fancy goods, laces and embroideries, lingerie, millinery goods, perfumery and soap, precious stones and imitations, veilings, and hides and skins. All but the last bear the peculiar stamp of French and especially Parisian industry.

268. Seaports. The greatest is Marseilles on the Mediterranean, having a population in 1906 of 517,000. The ground at the mouth of the Rhone is not suitable for a city ; hence the port has grown up on the coast to the eastward. The clearances of 1907 gave a tonnage of 5,879,000. The chief Atlantic ports are Bordeaux and Nantes. The wine production of the Bordeaux consular district in 1907 was 325,000,000 gallons and the exports of wine amounted to $23,000,000. This is the chief center of French foreign trade in wine. In 1907 the total export and import trade of Bordeaux with the United States was $13,000,000. It ranks second only to Havre in French transatlantic trade. As Bordeaux is the port of the Garonne, so Nantes is the port of the Loire, the two great river basins forming the hinterlands, while the Atlantic opens to the west.

Havre, at the mouth of the Seine, is the port of Paris and the second center of sea trade in the republic. It is the chief port for trade with America, — cotton, petroleum, tobacco, cereals, and other products being here entered. A new line of vessels has recently been established between this port and Montreal, and harbor improvements are planned, to receive the largest ships. Vessels of 5000 tons go up to Rouen and smaller craft follow the Seine to Paris and beyond. Atlantic liners call at Cherbourg and Boulogne, and Dieppe and Calais are French terminals of boats crossing the Channel to England. Dunkirk

FIG. 187. Marseilles and harbor

is on the Channel near its opening into the North Sea, and has grown rapidly in its shipping until it is now fourth in importance among French marine cities.

269. Foreign trade. The total in 1907 was $2,236,000,000, imports being about $100,000,000 in excess of exports. Like the United Kingdom, France brings in raw materials largely, but differs in the small imports of foodstuffs, as also of coal. Raw cotton is a heavy import from the United States, amounting to $55,000,000 in 1907. Raw copper was next, with $17,000,000; then machinery and tools, $11,000,000; and petroleum with its products, $7,000,000. Wood was the next item, with nearly $5,000,000. Exports of French manufactures to all countries in that year were valued at $629,000,000. France is not a country of the richest raw materials, and much of its prosperity is due to the spirit of its people, who make their land both attractive and productive. To the natural wealth which it has, "France adds the advantages of an industrious, frugal

FIG. 188. Exchange and wharves of Bordeaux

people, a scientific monetary system, and an army of artisans trained by education and inherited tradition to produce the highest artistic forms of manufacture, which compel the patronage of other nations." [1]

270. French colonies. The various colonies and dependencies occupy 4,000,000 square miles and contain over 50,000,000 people. They are mainly colonies of "exploitation," and are found in Asia, Africa, South America, and the islands of the

[1] Trade of France for the year 1907, *Consular Reports*, Annual Series, No. 13, p. 14.

sea. Their total trade with France amounts to about $250,000,-
000 per year. While this is important, it would seem that the
colonies add more to the prestige of the country than to its
commercial life. The home population has in recent times
varied little. Home resources are diligently used, and there
has been little colonizing, in the way in which men of the
United Kingdom have gone forth to other lands.

FIG. 189. Havre and the mouth of the Seine

Algeria and Tunis are the most closely related to France and
have the most trade with it. Wheat and other cereals are large
crops, as are grapes and tropical fruits. Much more than half
the colonial trade is with these countries of North Africa. Vast
areas of the western Sahara, the northern Kongo, and the
island of Madagascar are under French dominion. French Indo-
China lies on the China Sea, east of Siam, and has a population
of over 18,000,000. Its annual trade with France is $25,000,000
to $30,000,000. A few small islands and French Guiana alone
remain of the empire once held by France in the New World.

271. Belgium. Belgium is about one fourth larger than the state of Vermont and had an estimated population, at the close of 1907, of 7,317,000. This gives 643 persons to the square mile, probably equaled by no other country of Europe. Here is proof of the industrial and commercial capacity of a small land controlled by a diligent and skillful people. The country is here studied with France, because of close relations both of land

FIG. 190. Shipping in the harbor of Antwerp

and people. About half of the Belgians use French as their native tongue, the rest employing Flemish, which is related to the Dutch language. The highlands of the Ardennes continue from France into Belgium. They are rich in minerals in both countries, and the industrial part of Belgium is therefore continuous with similar territory in France. Iron and glass work are the special manufactures based on the minerals, and zinc is produced abundantly and is an important export.

Northern and western Belgium is a lowland adjoining Holland, and resembling that country in its surface and its carefully cultivated fields, yielding hay, cereals, and such special crops as sugar beets, chicory, flax, and hops. Besides the native minerals, imported diamonds make a great industry, with 5000 diamond cutters in Antwerp alone. So largely is the diamond market of the world centered in the United States that nearly one half of these workmen were idle during the financial depression in the United States in 1907, luxuries being the first to suffer when trade is slow.

Liege, in eastern Belgium, near the coal beds, is a center both for woolens and for metal work, particularly firearms and machinery. More centrally placed is Brussels, which, like other great capitals, is a home of general industry, lace making being the best known. The ancient Ghent, in the west, is a place of cotton manufacture ; and Ostend, on the sea, marks a point of departure for England. The great commercial city is Antwerp, on the Scheldt, whose estuary crosses a corner of Holland before Antwerp is reached. Antwerp is one of the ports of the world, dating from the seventh century. The total of entrances and clearances of ships for 1907 was more than 12,000, and Hamburg alone of continental ports surpassed Antwerp in tonnage in that year. Of regular steam navigation lines 110 engage in the Antwerp trade, and 9 of these ply to ports in the United States.

Belgium illustrates the importance of transit trade. Her general commerce for 1907 amounted to $2,183,000,000. Of this $1,000,000,000 represents transit trade. The special commerce of Belgium amounted to $1,177,000,000, a vast sum for so small a country. Imports for use in Belgium made a total of $661,000,000, and exports of Belgian products of farm, factory, and mine made $515,000,000. This points to industry, skill, and ample resources, but it should also be said that wages for common and for skilled labor are very low. The condition of the worker is inferior to that of his fellow in Canada, the United States, or Australia. And another result of low wages is the ability to manufacture at low cost and thus to offer formidable competition in foreign markets.

CHAPTER XXI

272. Central position. The commercial growth of a country depends upon its position, its natural resources, and the character of its people. In each of these respects the empire is favored. Seven countries border it, and all are progressive or have vast resources. The United Kingdom, Norway, and Sweden are distant but a short sail, and Italy and the Mediterranean are accessible by the passes of the Alps. Turkey, Greece, and the Iberian peninsula are the only parts of Europe more removed, and none of these has a first place in modern commerce.

This close relation will be clear if the surface of the land be studied. In the south is the northern border of the eastern Alps. North of this is the Bavarian highland, — around and north of Munich and drained to the eastward by the Danube, — which passes into Austria, giving an open road to Vienna, Budapest, and Constantinople. To the north nearly all of central Germany is upland, not of the plateau kind, but a region of many low mountain ranges and hills rarely too high or too rugged to be occupied by forests or highly tilled fields, and bearing innumerable villages and cities and supporting a dense population. Northern Germany is the Baltic lowland, to which Berlin is central. From the Baltic and the North Sea it stretches southward to Breslau, Dresden, Leipzig, Hanover, and Cologne, where it is replaced by the mountains of Bohemia and the central uplands of Germany.

This lowland opens the empire to the far-reaching plains of Russia on the east, and merges with the plains of Holland and Belgium on the west. Four great rivers drain northward large parts of the empire. Beginning in the west, these are the Rhine, the Elbe, the Oder, and the Vistula. The Rhine passes from the Swiss Alps through western Germany and enters the sea through Holland. It is bordered by a fertile land. Strassburg, Cologne, and Düsseldorf are its chief cities, and it leads down to the sea at the port of Rotterdam. The other three rivers rise beyond the boundary and flow across the German plain, — the Elbe into the North Sea, the Oder and the Vistula into the Baltic. Upon the Elbe is Dresden and on its estuary is Hamburg, greatest of continental ports. Breslau is the chief interior city of the Oder and Stettin is its port. The Vistula is less a German river than the others, but has Danzig at its mouth.

Berlin lies on the plain between the Elbe and the Oder and is joined to each by a canal. The southern plateau, the central highlands, the northern plain, and the five major rivers with their cities afford a framework upon which further knowledge of the country may well be placed. The climate of the west is like that of France, moderate in heat and cold, but the eastern or Baltic region has greater extremes and. is continental in its climate. The area is 208,780 square miles, about the same as that of France. The population in 1905 was a little more than 60,000,000 and the density was 290.

273. Forests and forestry. One fourth of the country is covered with forests, two thirds being coniferous, chiefly fir and pine. The forests are strictly controlled by law and receive assiduous care, the art of forestry standing here as a model for all nations. While in the United Kingdom forests cover but small areas and belong to a few, in Germany they are extensive and are so managed as to meet the daily needs of the people. No wood is wasted, and even private owners are not at liberty to cut timber except under wise restriction and with due provision for replanting. The remaining one third of German forests consists mainly of hard woods, among which oak and beech are most abundant.

274. Minerals and mining. The leading facts concerning German iron and coal have already been given in sects. 57 and 67. To the development of these minerals the empire largely owes its industrial growth and also its position as the second commercial nation of Europe. Silver, copper, zinc, and lead are the metals which, in addition to iron, are mined in considerable quantities. Deposits of salt abound beneath the northern plains. Stassfurt, which is in Prussian Saxony not far from Magdeburg, has not only rock salt but unique deposits of potassium and magnesium sulphates and chlorides and other salts, from which are derived many substances used in the arts. This small city has thus become one of the great centers of chemical industry. The mineral product of the empire for 1907 was valued at about $450,000,000. This may be compared with the product of the United Kingdom for the same year, $675,000,000, and with the output of the United States, $2,069,000,000.

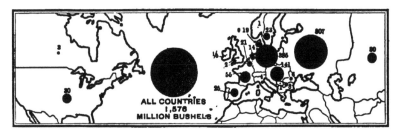

FIG. 191. Average annual production of rye in millions of bushels, for the years 1903–1907 inclusive

275. Agriculture. The productive land in Germany is considered to include 91 per cent of the whole. Remembering that land is required for roads, houses, parks, and various public purposes, it will be seen that the amount of waste land is small. The character of German farming will be well seen by comparing the areas under certain crops. A hectare is equivalent to 2.47 acres.

AREA OF CERTAIN GERMAN CROPS IN 1907

Rye	6,042,000 hectares
Hay	5,970,000 hectares
Oats	4,377,000 hectares
Potatoes	3,297,000 hectares
Barley	1,701,000 hectares
Wheat	1,746,000 hectares

The large amount of rye points to the fact that this is still used as a bread cereal by a great mass of the poorer people. In the relative use of wheat and rye Germany is in strong contrast with France, though it must be remembered that Germany imports considerable wheat. Another feature is the crop of potatoes, — more than is produced by any other country. The greater part of the land is in small farms. About six sevenths of the agricultural land is cultivated by the owners. Intensive culture of small plots by owners results in a large total product.

Besides the special features of German agriculture named above, two others must be given. One is the cultivation of the

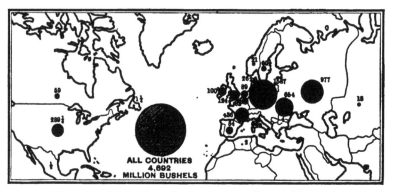

FIG. 192. Production of potatoes in millions of bushels

sugar beet, which is found chiefly on the northern plains. Under government bounties both breeding and culture of the sugar-producing beet have been brought to high perfection, and the empire provides its own sugar and exports large quantities. Thus dependence on tropical cane sugar is removed and a larger degree of self-support is gained by intelligence and industry. As potatoes and beets are among the distinctive crops of north Germany, so the vine is typical of the middle Rhine region in the south and west. Here the empire shares an industry which continues on a larger scale through Italy, France, and southwestern Europe in general.

276. Manufactures. Manufacturing industries in the empire have grown rapidly during the past thirty years. The number of people thus engaged has increased in proportion to the number tilling the soil. This does not mean abandonment of the soil, but rather the great growth of cities, the centers of industry and commerce. Of European nations Germany is second only to the United Kingdom in its manufactures, and is speedily coming to resemble it in making manufactures and trade primary, and in consequent importations of food and raw materials.

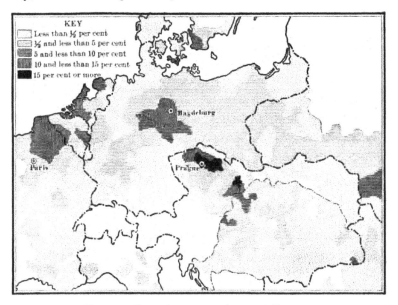

FIG. 193. Sugar-beet map of central Europe

The two leading groups in German industry are (*a*) iron-working, including machinery and instruments; (*b*) textiles. In making instruments of precision the thoroughness and the scientific training of the Germans find play. The textiles are of all the great kinds, — cotton, woolen, linen, and silk. For these all of the cotton and raw silk and much of the wool and flax must be imported.

Other kinds of manufacture may be noted as typical of the country. One of these, already named, is the making of beet sugar. Another is the making of chemicals, where again are seen both the high scientific training of the people and the varied mineral resources of the land. A third is woodwork, which employs several hundred thousand people. It is dependent on the abundant forests and reflects the patience and hereditary skill of the people. There is also high development in the working of precious metals and in the processes of printing and lithography.

277. Routes of trade. The railway system exceeds in mileage that of any other European country and is mainly under the management of the several German states. As with Paris and London, Berlin is the chief center of the network of lines, which reach thence to the chief ports and to all interior points. Viewed

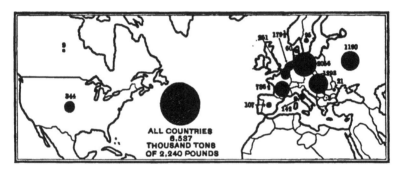

FIG. 194. Average annual production of beet sugar in thousands of tons of 2240 pounds, 1904–1905 to 1908–1909 inclusive

in their continental relations, several lines may be singled out. Those following the Rhine are not only the channel of home trade, but offer passage of men and goods from Great Britain via Rotterdam to Switzerland, the St. Gothard Tunnel, Italy, Egypt, and India. The express service from Paris to Constantinople crosses the Rhine at Strassburg and passes through Munich eastward. Likewise Cologne and points in France and Belgium find outlet through Berlin to Warsaw, to other Russian centers, and to the Siberian railway. From Hamburg lines run to Berlin and Dresden and through Bohemia to the heart of the Austrian Empire. The interior waterways are among the most highly developed in Europe. The five major rivers already named are improved for navigation and are joined to each other by canals, making 7500 miles of navigation, one fifth being by canals. North Germany is best supplied, owing to the greater depth of the rivers and to the ease of constructing

FIG. 195. The world's production of beet and cane sugar, for the years 1857-1858 to 1903-1904

canals across the plains. The Main is a branch of the Rhine, and a canal joins it to the Danube, so that a small craft may go from the North to the Black Sea. Another canal joins the Rhine to the Rhone. The greatest of German artificial water-ways, the Kaiser Wilhelm Canal, joins the North Sea to Kiel and the Baltic, and is of naval and commercial importance, making a saving of two days' time as compared with the passage to the north of the peninsula of Denmark.

278. Centers of trade. The greatest German port is Hamburg, on the estuary of the Elbe. At enormous cost the harbor has been fitted for large ships, and here is the German terminal for ships of the Hamburg-American Line. It is the natural Atlantic port for Berlin and much of the German interior, and is the greatest seaport of the continent of Europe. Next to

FIG. 196. Harbor of Hamburg

Hamburg is Bremen, on the estuary of the Weser, the terminal of the North German Lloyd Line. Excepting certain British lines, this and the Hamburg-American are unrivaled in the extent of their tonnage and the world-wide importance of their shipping. Stettin and Danzig are the chief Baltic ports. Stettin is near the mouth of the Oder and is a naval station and commercial mart. River and canal give to it a share in the commerce of Berlin and make it an outlet for much of the northern plains. Danzig is less important for Germany, but offers outlet for grain and lumber from Russian Poland.

Of interior centers the first is Berlin, the third of European cities and the sixth of the world. It has been a place of trade for several centuries, but has seen its greatest growth in the few decades during which it has been the capital of the empire. Here we add to London, New York, and Paris another city of the first order, which is a center of general rather than of special industry. About 150 miles southeast of Berlin, on the Oder, is Breslau, the capital of the province of Silesia, with 470,000 people. Here are exchanged the raw products of Russia for the more advanced products of Germany, and adjacent coal fields supply power for many industries.

Southward from Berlin, in the small but densely peopled kingdom of Saxony, are three great cities, Dresden, Leipzig, and Chemnitz. The first is on the Elbe, not far north of its passage through the border range of Bohemia. It has over half a million people and is a place of art and of artistic production. Near at hand are the porcelain works of Meissen. The second city is Leipzig, nearly as large, and also on the southern border of the German plain. It is the chief center of German books and printing, and the leading mart of the world for furs. In both of these lines of trade the United States is a large customer of Leipzig, but many Americans know the city best for its university and its music. Chemnitz has been called the Manchester of Saxony, and its quarter of a million people have chiefly to do with its machine shops and its mills for cotton, woolen, and linen.

On or near the Rhine are Düsseldorf, Cologne, Frankfurt am Main, and Strassburg. Düsseldorf, with a population of 250,000, is in close relation with the textile centers Barmen and Elberfeld. In the lower Rhine region are also Essen, the iron town, and Krefeld, second only to Lyons among cities of Europe for silks and velvets. Munich, the third city of the empire, is the capital of Bavaria. It stands on the plateau not far from the Alps, is a railway center and the headquarters of Bavarian trade. Largely through the influence of some of the kings of Bavaria, Munich is nobly built and has become a center of art and of artistic manufactures.

FIG. 197. Water routes and commercial centers of the German Empire

344

279. Foreign trade. The total exports and imports of the German Empire in 1907 amounted to about $4,000,000,000, placing the nation second only to the United Kingdom. Imports amounted to nearly $500,000,000 more than exports, the balance being made up by the income from foreign investments, the profits of German shipping, and the expenditures of foreigners in home territory. It will be remembered that French and British foreign trade illustrates the same conditions. In 1907 Germany imported breadstuffs to a value of $250,000,000 and the cotton importation amounted to $127,000,000. Other large items were wool ($94,000,000), raw silk, coal, copper, iron ore, rubber, hides, coffee, and eggs.

The exports form a larger number of items and consist mainly of manufactures. Coal ($62,000,000) was the only large export of raw material in 1907. Silk, woolen, and cotton goods, machinery, gold ware, glass, electrical materials, vehicles, aniline dyes and other chemicals were the leading items. The United States has larger trade relations with the empire than with any other nation except the United Kingdom.

280. German colonies and dependencies. With the exception of a small town on the coast of China, and a few islands in the Pacific, the German foreign domain is limited to west, southwest, and east Africa, the population of all dependencies being a little over 12,000,000. This number includes but a few thousand Germans; hence all are colonies of exploitation. The Germans who have emigrated have settled under foreign flags, particularly in the United States. The total annual trade between the empire and its dependencies amounts only to about $16,000,000.

281. Causes of growth. In 1833 the German states formed a *Zollverein*, or tariff union, which opened a free home market and gave a degree of common or national policy. Full union was achieved in the organization of the empire following the war with France in 1870. That war strengthened the unity of interest, added productive territory along the Rhine, and enriched the nation with $1,000,000,000 of war indemnity.

The Germans are descended from a hardy northern race and show vigor and persistent activity of body and mind. The birth rate is high and the population increased from 56,000,000 to 60,000,000 between 1900 and 1905. This increase has provided a large laboring class and has stimulated a great number to seek efficient training and higher standards of living. Thus leadership, skill, and expansion of markets have resulted.

Compulsory service in the army has given German youth strong discipline of body and mind, has widened their observation, made their national feeling strong, and promoted that obedience to law which is a marked character of the people, and which effectively aids good government, commercial honesty, and industrial progress. No country has organized so thorough a system of technical and commercial schools as Germany. Her factories, laboratories, and mines are filled with graduates of universities and higher technical schools, and men of the same training and efficiency sell her goods and promote her business in foreign lands. Contrary to the usual policy of trade in the United States, foreign representatives go equipped with the language of the country in which they are to work, and carry goods carefully adapted to local needs and tastes. Men of university training teach in common schools, and the direct influence of education in science is seen in industries such as textiles, mining, and metallurgy, the applications of electricity, and the making of a vast variety of chemicals.

With the consolidation of the empire, shipping has grown until Germany is second in the foreign carrying trade and has the two largest shipping companies in the world. Uniform tariff policy has been possible since 1870, for a single small state could no longer defeat the will of the majority. In 1879 Germany passed from a policy of virtual free trade to that of protection. In recent years she has added to her general tariff a treaty tariff policy, by which favors are given to countries favoring German trade, and specially high tariffs are put upon countries which discriminate against German products. With industry and commerce a general banking and financial system has grown, which is effective not only in the empire, but facilitates trade in many foreign centers.

The canals, railways, telegraphs, and telephones are chiefly under government ownership and are managed for the general welfare. The railways and canals work in harmony; each carries freight suited to the mode of transportation; and the railways cannot, as under private ownership, conspire to cripple water traffic. Secret rebates are not given to favored shippers, and industrial combinations are said to be comparatively free from stock watering and speculation.

Thus Germany in forty years has passed from the state of an agricultural people and become one of the industrial and commercial powers. With this change, as in the United States and Great Britain, has come migration to cities and the building up of great urban populations, joined to each other and to the outside world by every modern means of communication.

282. Germany and the United States. Raw materials must always constitute the bulk of German purchases from the United States. So long as the latter country has a surplus of raw cotton, copper, foodstuffs, and fertilizers, a German market is assured. The German effort to raise cotton in Togoland, in Africa, cannot be expected in any near future to cut off the demand for Southern cotton. Manufactured goods, however, will find a less certain sale. Often a temporary market is found, as for American machinery, but a German domestic product is soon forthcoming, built on the foreign model. It is as a competitor in manufactured goods that Germany compels the most serious attention of American industry and commerce. Her skill, cheap labor, shipping facilities, and her persistent promotion of foreign trade require equal skill and energy from those who would share the commerce of Latin America and the Orient.

283. The Netherlands. This country has about the same area as Belgium and is chiefly a low plain, continuous with the plains of northern Belgium and western Germany. Much of its surface belongs to the Rhine delta, and the several branches or "distributaries" of this river are the main avenues of transportation. Rotterdam is the principal port of the Rhine and ranks with Hamburg, Antwerp, London, and other seaports of the first order. Amsterdam is the largest city, being on the Zuyder Zee and communicating by canal and river with the North Sea and the Rhine. This country is mainly commercial rather than industrial. It has little coal and iron, and hence is different from all countries thus far studied in not having the basal materials of manufacture. It is open everywhere to the

FIG. 198. Harbor of Rotterdam

sea and to the interior of Europe, it has a vigorous people, and has therefore for centuries been one of the important shipping and trading nations of the world. Like all densely populated European lands, it imports much of its food and other necessities, but the tillage of its plains, often reclaimed from marshes and the sea itself, is intensive and agriculture shows some important exports. Among these are live animals, butter, cheese, eggs, flowers, bulbs, and vegetables. The vegetables often go to England and the bulbs are sent to all countries.

Three things may be observed : (1) the commercial rather than the industrial life, resulting from favorable position and water routes, with small natural resources; (2) that a great share.of. the trade is in goods which are in transit between the sea on the one hand and the German Empire and other countries on the other; (3) that the Netherlands has held, since the early modern discoveries, important and rich colonies with which there is a vast trade. Java alone of the East Indian possessions has about 30,000,000 people and is rich in almost every tropical product. The country has possessions in Sumatra, Borneo, Celebes, and other parts of the East Indies, and also in the Caribbean region. The chief exports of these colonies go to the Netherlands and are thence distributed to other lands, making up the greater part of Dutch foreign trade.

284. Denmark. Denmark is but little larger than Belgium or the Netherlands and has a much smaller population than either of those countries. Like the Netherlands, it has no important mineral resources. For fuel it must depend on peat, wood, and imported coal. Its merchant marine is considerable, but much inferior to that of the Netherlands. It has, however, been recently extended by regular sailings from Copenhagen to Baltimore and to Argentina. Copenhagen is the only large city, and the colonies, while somewhat extensive, are unimportant commercially; they are Iceland, Greenland, and a few small islands in the West Indies. The home country is preëminently agricultural, but the educational system is good, industry is carefully encouraged, the people are prosperous, and Copenhagen, with nearly a half million people, is one of the important cities of Europe. Oats, hay, and roots, including the sugar beet, are the chief crops, and dairying is the preëminent occupation. Butter is produced in the most sanitary and scientific manner, and the reputation of the Danish product is jealously guarded. Meat and eggs are large products, and horses are bred for export. The leading imports are coal, maize, oil cake and meal, lumber, iron, and steel, hardware and machinery being also important. The exports for 1907 made a total of $99,-000,000. Of this butter was credited with $43,000,000 and bacon and ham with nearly $26,000,000.

285. Norway. Norway is slightly larger than the United Kingdom, but has a smaller population than Denmark. Being a high and rocky country, agriculture is limited to the dairy, poultry, vegetables, and hardy grains, the whole coming short of feeding the population. The resources of the country are in its forests, its fisheries, and its shipping. Copper, lead, zinc, and iron are found, and the mining interest may increase. A

FIG. 199. Harbor and fish market of Bergen

vast amount of power now goes to waste in innumerable water-falls, and as electrical application of power has now been under-taken, future industries may be large. The Norwegians are sailors by tradition and training, and, for a small nation, have a remarkable share in the world's carrying trade, their fishermen, whalers, and merchant ships being found in all waters. About 15,000 craft are engaged in Norwegian fisheries, and the catch of 1907 was worth over $12,000,000. The forestry output, in-cluding wood pulp, was worth $25,000,000. The merchant marine contains nearly 2000 steamships and nearly 6000 sail-ing vessels. The broken shore of Norway with its islands and fiords favored the growth of a maritime people. Breadstuffs come principally from Russia and Germany. As with so many European countries, the balance of trade is against Norway; but it finds its compensation in the proceeds of shipping and of the tourist traffic, and in money sent by Norwegian Americans to their relatives at home. Fish and wood products (under this head come lumber, pulp, paper, matches) are the leading exports. Christiania and Bergen are the chief commercial and industrial cities.

286. Sweden. Sweden is about two fifths larger than Norway or the United Kingdom and contains a little over

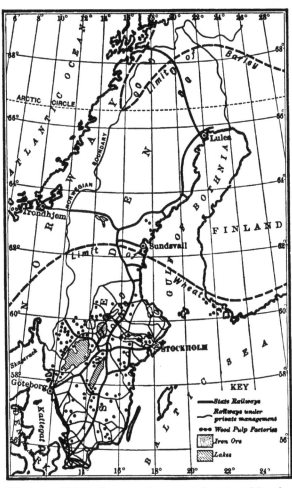

FIG. 200. Chief railways and certain industries of Sweden

5,000,000 people. It has much more plain and arable land than Norway, is equally abundant in forest, but does not from its position fill so large a place in the world's shipping. Like the United Kingdom and other countries of northwestern Europe, Sweden has a mild climate for its high latitude. It has important mineral resources, and thus in its variety resembles Belgium or the United Kingdom, but it is new in development as compared with either of these countries, and may be expected to make large advances in the coming years.

Railway building is favored by the abundance of wood and iron, and the mileage is over 8000, forming a network in the southern half of the kingdom and sending one line to the far north and a branch westward to Trondhjem in Norway. Barley, rye, and oats are the chief grains, potatoes are widely raised, and the dairy is so extensive that exports of butter to England are large. Manufactures of wood and iron furnish the largest industries, but others are growing by the enterprise of the people, and foreign trade is promoted by permanent exhibitions held in Cape Town, Valparaiso, Tokyo, Yokohama, and Hongkong. This activity is said to have led to the establishment of steamship lines to Chile, Japan, and South Africa ; and it shows how surely a vigorous people with rich natural resources takes its place in the commerce of nations. The coal supply is small, and limited to the south, but compensation is found in abundant wood, in peat supplies believed to be equivalent to 4,000,000,000 tons of coal, and in the available water power.

Stockholm is the largest city and is the capital, having 337,000 people in 1907. Göteborg is the second city, having, in 1907, 160,000 people. Being in the southwest, on the shore opposite Denmark, it is well placed for foreign trade. Both cities have large industries, among which are iron, cotton, and sugar. Great Britain and Germany hold much the larger share in the foreign trade of Sweden.

CHAPTER XXII

287. Switzerland, Italy, Spain, Portugal, and Greece. None of these countries is of the first commercial magnitude. Switzerland, though small, is one of the most advanced nations, and is thus closely related to France and Germany. It is also geographically joined to them quite as closely as to Italy. The other countries in this group are peninsular, and all but Portugal border the Mediterranean. Although neither Spain nor Italy nor Greece extends to 35 degrees north latitude, they all have in parts a subtropical climate, the productions differing from those of any European country thus far studied except southern France.

Italy, like Germany, was unified out of several smaller sovereignties during the nineteenth century, and is taking its place in modern commerce. Germany revives the commercial life of medieval and early modern centuries. Italy looks back not only to a medieval but to an ancient period, in which her peninsula was the commercial center of the world. Spain is rich in resources and is backward not by nature but through man. She once possessed vast colonies of exploitation and sluggishly relied on them. Shorn of these, with little shipping and small progress at home, Spain now begins to arouse herself to her needs and possibilities.

288. Switzerland. Switzerland is often considered as synonymous with the Alps. It must be remembered that the western Alps, including Mont Blanc, are in France, the eastern Alps are in Austria, and the southern Alps are largely in Italy. The central Alps are Swiss, and from their northern base a hilly upland stretches to the French and the German Jura. This upland and the Rhone valley are the chief regions of tillage.

A Swiss geographer has called his country an economic paradox, because it is so poor by nature and yet stands in the front

353

rank of European foreign commerce in proportion to its population. For this result several reasons may be given.

1. Switzerland is poor in minerals but abundant water power more and more takes the place of coal. The latest figures show over 200,000 horse power used and ten times that amount in reserve. It would be possible to operate the entire Swiss railroad system and still have a considerable surplus for industry.

FIG. 201. Montreux and vineyards, Lake Geneva

2. The frugality, industry, and intelligence of the people preserve and increase the national wealth, and skill in various industries has accumulated through generations of experience.

3. Excellent markets are found among the populations of several adjacent countries, particularly France and Germany, and even in a land so remote as the United States.

4. These markets are accessible by good roads, in whose building the Swiss people have had a full share. By advanced road building they also make their country one of the chief channels of trade between the commercial nations in the north and the entire Mediterranean region and the Orient.

5. Nature has endowed Switzerland with attractions of scenery and made it a field for rest and sport, which attracts travelers of all nations and swells the national income.

Milk products, especially cheeses, are the chief substances belonging to agriculture which figure in the export trade. The greater industries are in cotton, silk, and watchmaking. Both in silk and watches the trade with the United States has declined

FIG. 202. Spiral tunnels of St. Gothard railway, Switzerland

because of the growth of these industries in the latter country. The cotton trade between the two countries, however, continues large. Like Germany, Great Britain, or Italy, Switzerland imports most of her raw cotton from the United States and sends to us half of the Swiss output of embroideries, or a value of nearly $18,000,000 in 1907. Embroidery may be called a typical Swiss industry, employing 6000 power machines and 16,000 hand machines, while the United States has of both but 616 machines. Zurich, Basel, and Geneva are the largest centers of industry, while Lucerne, Interlaken, Geneva, and many smaller places attract the traveler.

The chief routes of transportation are as follows :

1. By the St. Gothard railway and tunnel, from Brussels, Antwerp, and Rotterdam, following the Rhine or coming to it

FIG. 203. Chief routes, centers, and industries of Italy

at Basel ; thence to Lucerne and by the tunnel to Milan and the south. Traffic from Berlin and eastern Germany and from Munich centers likewise upon the St. Gothard Tunnel.

2. From Paris to northern Switzerland and Italy. Railways cross eastern France and pass through the Jura, and one line goes between the Jura and the Vosges, all centering upon the St. Gothard. The Loetschburg Tunnel, now under construction, will lead directly from Bern to the Simplon Tunnel, through the Bernese Alps, and give a more direct passage from Paris to Italy.

FIG. 204. Southern entrance to the Simplon Tunnel, Iselle, Italy

3. From Paris to Geneva, and thence up the Rhone valley to Brieg and the Simplon Tunnel, leading down to Milan. Of these two great Swiss tunnels, the St. Gothard is over nine miles long and was first opened; and the Simplon is over twelve miles long, is much nearer sea level, and was opened in 1906.

289. The Italian land. Broadly based on the Alps, the peninsula extends to the southeast 700 miles, almost bisecting the Mediterranean. In the north the plains of the Po continue from Turin past Milan to Venice and the Adriatic shore.

North and west are the Alps, which south of Turin curve to the east and pass into the Apennines. This Italian range rises like a wall on the south border of the northern plain, and thence southeastward nearly fills the peninsula. Italy communicates with nations of north-central Europe: (1) along the Riviera between Genoa and Marseilles; (2) by the Mont Cenis Tunnel route from Turin to Paris; (3) by the Simplon and

FIG. 205. Drying macaroni, Amalfi, Italy

St. Gothard routes from Genoa and Milan; (4) by passes across the eastern Alps into Austria; (5) by a coast route around the head of the Adriatic to Trieste.

Lines of railway border the Po plain at the base of the Alps from Turin to Venice, and at the base of the Apennines (the Emilian Way of the Romans) through Parma and Bologna. They also follow the east and west coasts to the end of the peninsula, and cross the Apennines at many points from sea to sea. The Apennines are so rugged that there are numerous tunnels not only on the crosslines but in parts of the coastal routes.

The forests are poor, much mountain ground is denuded and
washed, and the prevalence of goat keeping is often unfavorable
to the renewal of forest growth. Mineral resources are small.
There is little coal, and this fuel comes from Germany by the
St. Gothard and from Great Britain by sea. Petroleum must be
had from the United States and Russia. There is some zinc,
lead, and quicksilver, a little iron, and the marble of Carrara is
well known. Sulphur, which is largely mined in Sicily, is the
most important native mineral. The entire mineral output, exclud-
ing quarry products, in 1907 was worth less than $18,000,000.

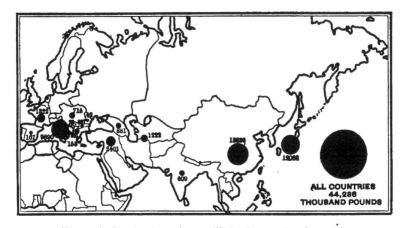

FIG. 206. Production of raw silk in thousands of pounds

290. Italian agriculture. This is the prevailing industry, and
land which would remain wild in many countries is improved
by fertilizing and laborious terracing. Wheat is the first of
Italian cereals, and maize is also much raised for human food,
the warm climate being suited to its culture. More rice is raised
than in any other European land. Vegetables are an important
share of the food, and help to show how a large population
(over 30,000,000) can live in a country of small resources.
The climate, however, is mild, little fuel is needed, and the
standards of living for the poorer classes are low.

The characteristic products of Italian climate and soil are
grapes and wine; olives and olive oil; lemons, oranges, and

FIG. 207. Genoa and its harbor

mandarins; figs, nuts, and silk. Wines are annually made to the value of $200,000,000, of which much is exported. The mulberry tree is a common feature of the landscape, its leaves serving as food for the silkworms. China and Japan alone surpass Italy in the amount of raw silk produced. The climate is too warm for the potato to do well. Hemp, flax, and cotton are grown to some extent, but the raw cotton is chiefly imported. Irrigation is much practiced, modern machinery is introduced on the northern plains, and the culture, whether carried on by antiquated or by modern methods, is so intensive as to awaken admiration. It demonstrates that land need not be "run out" or unproductive even after it has been under crops for two thousand years.

291. Italian cities. Milan and Turin are the great cities of the northern plain, and the former, in the center of rich lands and fed by the routes across the Alps and the streams of trade from Genoa and the sea, is the most important commercial center of the kingdom. On the east, Venice represents the commerce of the Middle Ages, and in the west, Genoa is the first Italian seaport of to-day. Leghorn, on the west coast, has taken commercially the place of Pisa, also of medieval renown, and has a highly important foreign trade. Naples is the largest Italian city, but is largely a place of call for passenger traffic. Palermo, the port of Sicily, is an Italian city of the first rank, while Rome and Florence are centers of varied trade, but find their chief interest in matters of art, history, and government.

292. Italian foreign trade. The total for 1907 was about $900,000,000, the imports as usual largely exceeding the exports. This foreign business is the result of vigorous development almost confined to a single generation, and is the largest example of the new life of the Mediterranean region. The Italian peninsula is central and therefore well situated for both oriental and western trade. Suez, Gibraltar, and the Alpine passes are the outlets to regions beyond the Mediterranean, and these are used by growing railway traffic and a rapidly increasing merchant marine. The tourist traffic from America

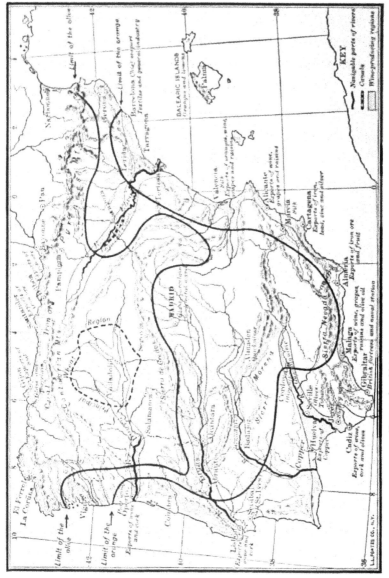

FIG. 208. Chief routes, centers, and industries of Spain

to Naples and Genoa direct is large and is conducted by English and German steamship lines. Italian steamships ply to the United States, Brazil, Argentina; and in the East to Egypt, Syria, and Constantinople. Commerce with both Americas is stimulated by the emigration of millions of Italians to the United States and Argentina, for these people desire some products of their native land. The raw-cotton import from the United States in 1907 was nearly $40,000,000, and this illustrates the keen industrial life which is springing up in the progressive parts of Italy, especially in such cities as Milan and Turin. More than 5,000,000 cotton spindles are now at work in Italy. Goods worth $40,000,000 were landed in Genoa from America in a single recent year. The Italian government has promoted trade by shipping subsidies, by extending the railroads, and by taking part in American and other foreign expositions.

293. Spain. This country is shut off from the continent by the Pyrenees, which, though not so high as the Alps, are a more complete barrier. Railroads enter Spain at the ends of the range. Except on the side of Portugal, Spain is elsewhere washed by the sea. Ports are few and the rivers for the greater part are not navigable, presenting conditions opposite to those of the United Kingdom or the German Empire. Still the isolation of Spain must be ascribed more to the spirit and history of the people than to the facts of her physical geography. Two thirds of the people can neither read nor write, and until recently there has been more reliance on the wealth of colonies than on the use of home resources. A standing army of 100,000 is maintained. Methods are backward and taxes are heavy.

In natural resources perhaps no European country is better endowed. The mineral wealth is great and in remarkable variety, including coal, iron, copper, zinc, cobalt, lead, mercury, sulphur, gold, and silver. There is coal in many provinces, but it is little developed and much is imported, especially from Great Britain, in ships which carry back iron ore from the Bilboan region. Bilboa is the center of this traffic and has also begun the making of iron and steel.

The agriculture will be best understood if it is remembered that central Spain is a plateau of more than 2500 feet of altitude, crossed by east-and-west mountain ranges. It is lacking in rainfall, is cold in winter, hot in summer, and thinly peopled. Railway communication is difficult and waterways are impossible. Between this plateau and the Pyrenees are important lowlands. In the basin of the Douro, the region around Valladolid, is the most important wheat region, and this cereal nearly meets the needs of the people, though some is imported. To

FIG. 209. Harbor of Barcelona

the east the basin of the Ebro is agriculturally productive, as is the coastal border along the Mediterranean east and south. As the north has a rich lowland, so has the south, in the province of Andalusia. These lands lie in the valley of the Guadalquivir, south of the central plateau and north of the lofty and snow-clad Sierra Nevada range.

Barcelona, well to the north, is the chief Mediterranean seaport of Spain, and is the center of the industrial region of the province of Catalonia. Farther down the coast is Valencia, the center of the orange trade, and thence are shipped many onions, a crop whose total value for the country is from

$10,000,000 to $20,000,000 per year. In this region irrigation
is practiced and several crops are raised each year. Irrigation has
long been used, and in much of the country is essential to production. Mineral industries are tributary to Almeria, a port of
the southeast, and to Huelva, a port on the Atlantic. The
ancient Cadiz is of declining importance. Wine, fruits, nuts,
and raw ores of iron, copper, and other metals are the larger

FIG. 210. Wharves at Oporto

items of export. Quite unlike the advanced countries of Europe,
this venerable nation exports raw products and imports manufactures, and yet there is, as stated, coal; and there is in many
provinces a considerable amount of water power that could be
used. In 1907 the total foreign trade was $310,000,000, equally
divided between imports and exports. Spain has an area of about
190,000 square miles and a population of nearly 20,000,000. We
may compare it with Switzerland, having an area of less than 16,000
square miles, with small natural wealth, with less than 4,000,000
people, and having a total foreign trade of $564,000,000.

Spain is, however, awakening; efforts are made towards higher standards of education, industry is slowly rooting itself, and this backward people, with a country that is rich and varied in resources, will in time take its proper place in the work and trade of the modern world.

294. Portugal. This country occupies 34,345 square miles, or a little more than one seventh of the Iberian Peninsula, and contains about 5,000,000 people. All that has been said of the

FIG. 211. Wharf and shipping of Piræus

backwardness of Spain is true of Portugal. Its glory is in the explorations and colonial extension of the past, though it still retains the Azores, Madeira, and important territories in Africa. It is mainly agricultural, but imports a part of its food-stuffs and most of its manufactures, though there is, as in Spain, some awakening of industry. Its largest trade is with Great Britain. It imports from the United States chiefly corn, cotton manufactures, petroleum, stoves, and wheat. Its part in the world's trade is small, the total foreign commerce amounting annually to about $100,000,000, the imports being double the exports. The resources of the country are increased by the return of persons of wealth from Portuguese colonies and from Brazil, to take up residence, especially in Lisbon. This city, the capital of the republic, and Oporto, are the chief seaports; the latter is the seat of the wine trade, wine and cork being characteristic exports.

295. Greece. The area is about 25,000 square miles and the population is between 2,000,000 and 3,000,000. This ancient

FIG. 212. The Corinthian Canal

land feels also the new impulse of modern life. In spite of emigration the population has quadrupled in the past seventy-five years. Electric tramcars ply in Athens and the Piræus, and surveys are being made for irrigation works. During a single recent year 277 building permits were issued in Athens, and dry-docks costing $1,000,000 are under construction at Piræus, the port of Athens. In several mills millstones have been replaced by modern milling machinery brought from England. Automobiles for freight and passengers are in use in Athens and in Sparta. The total foreign trade of Greece is about $50,000,000 per year. The mineral output is considerable, including Pentelic marbles exported to New York and other parts of the United States. Fruit, olives, tobacco, and wine are among the principal items of the export trade. The Corinthian Canal has not become a great highway of commerce, but is of much interest because it marks the realization of a project cherished since the classic period. Greece shares in large measure the awakening which is seen in southern Europe. It is most conspicuous in Italy and least marked in the countries of the Iberian peninsula. The Greek immigrant has become almost as well known to Americans as the Italian, and his capacity for trade is seen not only in the great cities but in many of the smaller towns of the United States.

CHAPTER XXIII

EASTERN EUROPE

296. General survey. Here are included Austria-Hungary, European Russia, and several small countries in the Balkan region and on the lower Danube. In several respects this part of Europe may be called continental: (1) in extent, for considerably more than half of the area of Europe is embraced in the countries named; (2) in its relations, for it is joined to Asia by a long land boundary, and its waters are either inland, arctic, or comparatively closed, as the Baltic, Adriatic, and Black seas, — these waters being somewhat isolated from the greatest ocean routes; (3) in climate, since open oceans are remote, their ameliorating influence is not felt, and cold winters and hot summers prevail. Still the rainfall is nearly everywhere enough for crops.

This part of Europe is partly Asiatic in its types of men. In the west, however, especially in Austria, the German branch of the Teutonic family prevails, and in the parts bordering Italy and even on the lower Danube are people of Latin origin, offering a mixture of races and languages unknown farther west. The northern parts are flat and undisturbed; the southern parts are broken into mountains of many ranges, all belonging to the great system which extends from the Atlantic Ocean through southern Europe to eastern Asia. Agriculture is the ruling occupation, Austria alone belonging largely to the industrial side of Europe.

297. Austria-Hungary and the Danube. The river is the best key to these countries. It has already been described as crossing the plateau of Bavaria. At Passau it passes into Austria, and a short distance below Vienna, at Pressburg, it becomes a Hungarian stream. Approaching the heart of Hungary, it

369

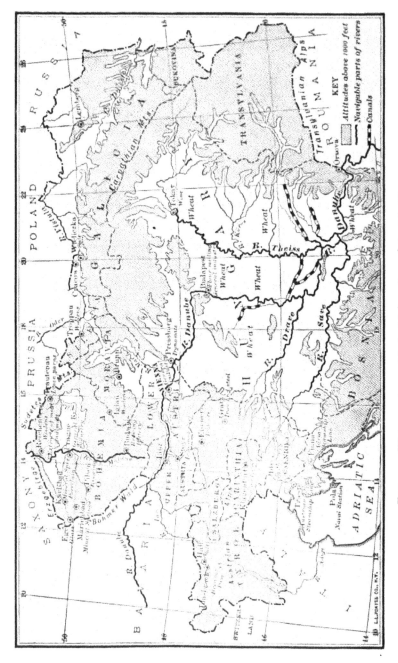

FIG. 213. Water routes and industrial centers of Austria-Hungary

turns to the south past Budapest, and in southern Hungary it receives the Drave from the west and turns eastward, leaving the empire at Orsova, through the Iron Gate. It is the second river of Europe and the main artery of the empire. The government has improved navigation at Vienna, and at the Iron Gate one of the channels entering the Black Sea has been deepened by jetties, and the navigation is free to all nations. The Ludwigs-Kanal in Bavaria joins the Danube to the Main and the Rhine.

Along the Danube runs the great route from western Europe to Constantinople, and it would have even higher importance if the Black Sea were nearer the centers of trade. Germany and the Rhine, or the United States and the Mississippi, offer natural comparisons, but in neither case is the river so important to the country as the Danube is to this empire. The two capitals are upon its banks, fertile plains border it, and great rivers flow into it.

298. Austria. The provinces of this kingdom will be best understood if Vienna is taken as the point of view. It is at the gateway where the Danube passes between the west end of the Carpathians and the east end of the Alps. In fact, Vienna is at the northeastern corner of the Alpine territory which belongs to Austria. Thence Austria stretches west 300 miles to the border of Switzerland and south to the Italian boundary. Tyrol is the western division, and across it runs one of the great commercial routes from central to southern Europe. It passes from Bavaria up the Inn valley, over the Brenner Pass, and down the Adige to the plains of north Italy. We recall the St. Gothard and Simplon routes in Switzerland and the Mont Cenis route from France, making the four great international highways across the Alps. On the southern slopes of the Austrian Alps wine and silk link Austrian industry to Italy. At Eisenerz in Styria is a very important mass of iron ore, worked since Roman times, and much of it is carried northward down the valley of the Enns to the iron town Steyr, not far from the Danube.

Southward from Vienna, by the Semmering Pass, a railway crosses a mountainous country to Trieste, the only important seaport of Austria. It is 367 miles from Vienna, has about 200,000 people, and its commerce is large and is growing under the fostering care of the government. Its trade is over $400,000,000 per year, about equally divided between imports and exports, textiles and raw cottons being the largest items.[1] Over 14,000 vessels make up the Austrian merchant marine.

FIG. 214. Harbor of Trieste

Northward from Vienna, Bohemia is the most important part of Austria. Its capital is Prague, with 500,000 people, on the Moldau, a branch of the Elbe. Bohemia is the basin of the upper Elbe, and is walled in from Germany on the west and north by low mountains, a part of which is called the Erzgebirge, because full of useful minerals. Bohemia has ample coal, is densely populated, and is a most productive industrial region.

[1] Austria-Hungary, Trade for the Year 1907, in *Consular Reports*, Annual Series No. 19, p. 17.

Its communications are largely southward to Vienna, and northward following the Elbe through the mountains to Dresden and Hamburg. Mining, textiles, Bohemian glass, the beer of Pilsen, iron, and steel all represent large industries and make Bohemia the chief manufacturing region of Austria. In the west are Karlsbad and Marienbad, known among the watering places of Europe. More than one third of the population of Bohemia is German.

FIG. 215. The Danube Canal, Vienna

Eastward from Bohemia and north of the Carpathians and of Hungary, Austria includes Moravia, Silesia, Galicia, and Bukowina ; the last two a part of the plain continuing through Russia. Through the first two runs one of the chief routes of Austria, from Vienna to the valley of the Oder and eastern Germany. In Galicia, at Wieliczka, near Cracow, are salt mines of high value and widely known. They have been worked for 600 years, and the beds are said to be 500 miles long, 20 miles wide, and 1200 feet thick. These and other salt deposits in Austria-Hungary and Roumania offer inexhaustible supplies.

299. Vienna. This is the gate city between central and southeastern Europe, and its routes lead east and west, south to the Adriatic, and north by the Elbe and the Oder. It is therefore at the crossing of great ways, surrounded by varied resources, and is the capital of the empire. Like most of the larger capitals, it is a center of varied industry, as of art and education, and contains 2,000,000 people, thus being one of the seven largest cities of the world.

FIG. 216. Budapest

300. Hungary. This is one of the most compact countries of Europe, rimmed by mountains and traversed by a master river. Beginning at Vienna the long curve of the Carpathians shuts it in on the northwest, north, east, and southeast, coming down to the Danube again at the Iron Gate. Stretching broadly are the plains of Hungary, smooth and fertile as the prairies of the Mississippi Valley. It is an agricultural country and is more fully such than any European country yet studied, excepting perhaps Denmark. Its chief cultures are in wheat, barley, oats, rye, maize, potatoes, sugar-beet root, fodder-beet root, and grapes.

The raising and export of animals is also of high importance, and the forests and timber interests are large. The mineral resources are rich and include coal and iron.

The government vigorously aids manufacturing, granting low rates of transportation for machinery, offering subsidies and freedom from taxation to certain industries, and providing for technical education. The improvement of the rivers is undertaken,

FIG. 217. Harbor of Fiume

and canals are projected for joining the greater streams and thus shortening routes. There are 12,000 miles of railway, the main lines centering in Budapest, a splendid capital standing on both banks of the Danube and containing over 700,000 people. Its largest industry is the manufacture of flour, for which the city has the highest appliances and an international reputation, being well placed in reference to wheat fields and markets.

In the southwest Slavonia and Croatia belong to Hungary and give access to the Adriatic and to Fiume, which is the port of Budapest as Trieste is the port of Vienna.

301. Foreign trade of the empire. The population is about 45,000,000 and the total foreign trade in 1907 was $924,000,-000, imports and exports being about equal. This total scarcely varies from that of Italy. There is, however, a vast trade between the two kingdoms of the empire, which does not appear in this figure. The same is true in the United States, where the trade among states is several times greater than the foreign trade. Hungary has fully two thirds of its trade with Austria, and the two countries supplement each other, one being agricultural, the other largely industrial. Hungarian flour is sold widely in the markets of central and western Europe. The largest single import of the empire is raw cotton, and about two thirds of this is from the United States. Much of it comes by way of Bremen, but much also by Trieste, and the latter route is favored by the authorities. Among imports from America are character- istic mechanisms such as tools, typewriters, cash registers, and sewing machines. Through kinship and a common boundary the empire has a heavy trade with Germany.

302. Roumania. This country has 50,720 square miles,[1] being a little larger than the state of New York. It extends along the Danube from the Iron Gate to the Black Sea, having the Car- pathians and Hungary on the northwest and the plains of Russia on the northeast. Except on the Carpathian slopes it is a plain country, and like Hungary is agricultural. The population is about 6,000,000, and Bucharest, the capital, has more than a quarter of a million. Salt, coal, and petroleum are considerable products. One hundred thirty million tons of coal were mined in a recent year, and salt and petroleum furnish noteworthy exports. Wheat and corn are large crops grown on the plains of the lower Danube, and Roumania is one of the important wheat-exporting countries.

Belgium, Italy, Holland, and the United Kingdom take most of the exports, Belgium alone taking about $30,000,000 in 1906, of which nearly $28,000,000 was in cereals. The largest im- ports are from Austria-Hungary, Germany, and the United

[1] Statesman's Year-Book, 1909.

Kingdom, textiles and metal manufactures being the leading items. The total foreign trade in 1906 was $176,000,000, the balance of trade being in favor of the home country.

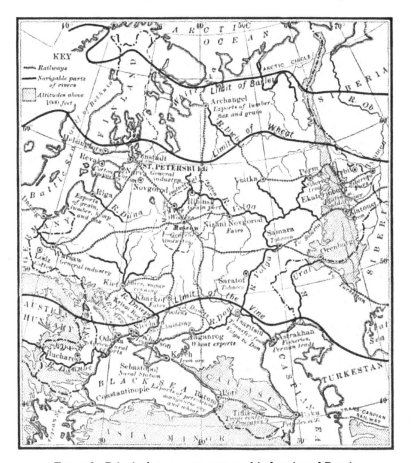

FIG. 218. Principal routes, centers, and industries of Russia

303. European Russia. The industries and trade of Russia are in sharp contrast to those of western Europe. European Russia embraces almost half the continent, and had, in 1907, 111,000,000 people, a little more than a quarter of the European population. Viewed in relation to the Russian Empire, European Russia, great as it is, has less than one fourth the area, but contains nearly three fourths of the population, so vast is Asiatic Russia and so scattered are its people.

Russia may be compared with the central plains of Canada. Both have barren tundras in the polar territory; in each is a great belt of forest as one goes southward; and in both the forests are succeeded by vast plains of fertile soil chiefly devoted to tillage. The forest belt has coniferous trees in the north and deciduous growth on its southern side. About 40 per cent of the country is in heavy forest, and, considering the vast area, this points to large resources in timber. About two thirds of this land is state property, and its development will be slow.

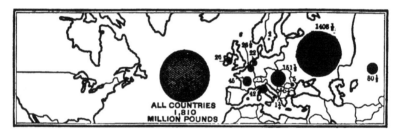

FIG. 219. Flax fiber; average annual production in millions of pounds, 1903–1907 inclusive

Russia is, next to the United States, the greatest producer of petroleum, and the fields of Baku on the Caspian Sea offer a unique example of a vast product from a small area. Other mineral resources are great, but are imperfectly developed. Iron occurs in Russian Poland, in the vicinity of Moscow, in the Urals, and in the basin of the Donets. This river is a branch of the Don, in southern Russia, in the region of the industrial and trading city of Kharkov. Here and in Poland coal is found, and the conditions for iron manufacture are present. Industrial progress is shown in the protection extended by government to domestic work in coal and iron, and in the establishment in the Donets region of a life-saving school to provide for safety in mining. Platinum should be named here because nine tenths of the world's supply of this rare but important metal comes from the European slope of the Urals. London and Paris firms buy nearly all that is mined, and thus are able to hold a monopoly and fix prices. Salt occurs largely, and the Galician deposits of Austria lie close to the Russian border.

304. Russian agriculture. Large as the arable lands are, they are limited by extensive tundras and forests in the north, swamps in the west, and salt steppes in the southeast. We may picture the agriculture by placing oats, rye, and barley in the

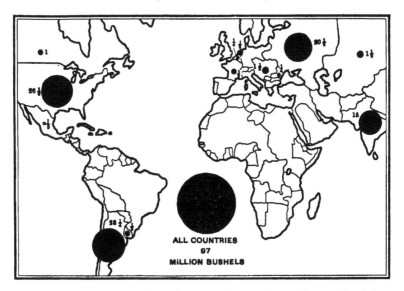

ALL COUNTRIES
97
MILLION BUSHELS

FIG. 220. Average annual production of flaxseed in millions of bushels,
1905–1907

north, wheat on the central and southern plains, and rice and vines in the south. Rye is more largely grown than wheat, because it is the home bread crop. Wheat, however, is better known, because it is the export crop. Poland and the black-earth region of central and southern Russia devote 1,000,000 acres to sugar beets, supplying all the home needs and exporting largely. Hemp and flax are grown in the Polish and Baltic districts and in central Russia, Russia leading all other lands in these crops.

Potatoes are much grown, and by considering Russia, Scandinavia, Germany, the United Kingdom, and other lands it is

Fig. 221. Harbor of Odessa

again seen that the fortunate finding of this plant in the New World has largely changed the food supply of Europe. The live-stock interests of Russia are superior to those of any other European country. Vast as Russian agriculture is, the methods are primitive, and the improvements now begun may be expected to cause great growth of products in the future.

305. Cities and the routes of trade. We begin in the central region, where the plain is a few hundred feet higher than on the borders. Here the great rivers rise, the Volga flowing to the Caspian Sea, the Don and the Dnieper to the Black Sea, the Düna to the Baltic, and the Dvina to the Arctic Ocean. Astrakhan is at the mouth of the Volga; Taganrog, the wheat market, at the mouth of the Don on the Sea of Azov; Odessa, near the entrance of the Dnieper upon the Black Sea; while Riga is the port of the Düna, and Archangel of the Dvina. There is fishing trade at Astrakhan, and Odessa deals in the cereals of southern Russia and has flour mills, sugar refineries, and many other industries. Cereals, flax, and lumber go out from Riga, and cotton, coal, and machinery come in, and there are many industries. Lumber and flax are exported from Archangel.

The rivers named and many others are navigable, offering over 16,000 miles for steamships, 8000 more for small sailing vessels, and 26,000 miles for rafts. These waters lose part of their value because of ice in the winter both in the streams and at the ports to which they lead.

We may now view the land traffic, and here we begin also in the central region, at Moscow, with a population of 1,350,000, the chief center of industry in Russia. Not only are there many factories, but house industries are carried on throughout the region. Not far south of Moscow are fields of coal and iron. Moscow is the center of the railways. The chief lines run as follows: westward to Riga; northwest to St. Petersburg; north to Archangel; eastward to Siberia and the Pacific coast; southward to Kharkov, Kief, and Black Sea ports; and west-southwest to Warsaw, Germany, and western Europe. From the Baku oil region on the Caspian Sea a road runs to ports on the Black Sea.

Warsaw, the third city of Russia, with 750,000 people, is the industrial and commercial center of Poland. Lodz, also in Poland, is a textile center and is nearly half as large as Warsaw.

Nizhni Novgorod, east of Moscow, is the seat of the greatest of Russian fairs, formerly the great means of the interior trade of the empire, and even yet transacting business to the extent of many millions of dollars. St. Petersburg, last to be mentioned, is the first city of Russia, has about 1,700,000 people, is second only to Moscow in industries, has a large foreign trade, and is the financial center and the capital of the empire.

306. Foreign trade. The foreign commerce amounts to about $1,000,000,000. This is little more than that of Austria-Hungary or Italy, but Russia is the awakening giant of Europe. It must also be said that the internal trade is vast, and that no other nation of Europe has resources that may make it so nearly self-sustaining as Russia. Russia buys in foods chiefly tea, coffee, wine, and fish. Other imports are rubber, coal, cotton, wool, chemicals, stationery, and machinery. She sells chiefly breadstuffs, butter, eggs, flax, hemp, hides, oil cake, petroleum, and timber. Much the greatest trade is with Germany, Great Britain and the United States standing second and third in Russian commerce.

307. Southeastern Europe. The Balkan region toward the Adriatic belongs to Austria-Hungary ; and Greece, the southern part of the peninsula, has been studied with southern Europe. There remain Bulgaria, Servia, Montenegro, and Turkey in Europe. Much of the region is rugged and barren, but the lowlands are genial and fertile, and, but for the misuse of man, might be of importance at home and abroad. Bulgaria and Servia raise wheat, and indeed rank among the countries of eastern Europe which export this cereal. Both of these countries have also the advantage of the Danube on the north as an artery of communication, and the valley of the Morava offers an easy route through Servia toward the important port of Salonica. Silk, tobacco, and wine are products of Bosnia and Servia, and minerals are abundant, but not much developed in the mountainous lands of the Balkan Peninsula. These regions play a small part in international trade, and for the purposes of this book may be passed with brief mention. Even Constantinople, city of the ages though it is, the gateway of Europe and Asia, has small industrial significance, and its commerce has declined in recent times because the two continents have opened other avenues, — through Russia in the north and by the Mediterranean and Suez in the south. Recent political changes, however, are introducing more modern and liberal conditions. Shipping from all ports on the Black Sea must pass through the Bosporus, and the city may again gain in the world of commerce a place suitable to its antiquity and its fame.

CHAPTER XXIV

ASIA

308. The continent as a whole. Asia has about four times the area of Europe, and contains more than twice as many people, yet its entire foreign trade is much less than that of the United Kingdom, which is smaller than Japan. Europe has a central lowland and its mountain barriers are toward the Arctic and Mediterranean seas. Asia has a central highland of plateaus and mountains, and its major rivers flow outward to the Arctic, Pacific, and Indian oceans. The form of Europe gives it commercial unity. Its longest railroads cross its central parts, and many of its greatest ships anchor almost at the heart of the continent. In Asia the high central areas are mainly unproductive, and trails and caravans form the only means of communication. The productive countries are on the edge of the continent and the chief communication is by sea. The commerce of Europe might be called centripetal, while that of Asia is centrifugal.

Much of western and central Asia is desert because removed from the ocean. This remoteness is due both to the size of the continent and to the presence of Europe and Africa between it and the Atlantic. The southeast and east are within the belt of the monsoons of the Indian Ocean, and both temperature and rainfall aid in making these parts productive. This part of Asia has been exploited by Europeans since ancient times for its spices, silks, dyes, and precious stones, and now affords European nations a commerce much more varied and capable of great growth.

309. Types of commercial development. At least five regions may be named.

1. Turkey in Asia includes some of the most ancient seats of commerce, — in Mesopotamia and about the eastern shore

of the Mediterranean. After prolonged desolation these regions begin to revive and to adopt the ways of modern times.

2. India and the East Indian islands are regions of tropical climate, of metallic and plant wealth, where the natives have little initiative and are ruled by European and American powers. Under such rule development has been large, and we have the most important examples of colonies of exploitation.

3. China has a larger population than any other nation and represents one of the most ancient civilizations. It is tied to custom and slow to move, but is beginning to learn the vastness of its resources and to awaken to industry and to international trade.

4. Japan is the only oriental nation of great initiative, swiftly learning its lessons from western countries and now taking its own part in industry and in the world's trade.

5. Siberia represents the expansion of Russia eastward during modern centuries. The resources are limited by the high latitudes, but still are large. The country is now opened by one trunk line of communication. It is the chief example in Asia of a new or frontier land, comparable with lands in North America during the period of exploration and westward movement.

Asiatic commerce, ancient and modern, is shaped by physical conditions, but even more by the spirit and traditions of its people.

310. Turkey in Asia. Following the Balkan parts of Europe, it is natural to take the Asiatic provinces of the Turkish empire. In this region one is ever reminded of ancient literature and history. Smyrna, Tarsus, Antioch, and Damascus, Palestine, Arabia, and the empires of Mesopotamia are brought again to view. .

Smyrna is the chief port and city of Asia Minor, having 200,000 people. Over 5000 vessels entered here in a recent year. Figs, licorice, olive oil, opium, tobacco, and emery stones are among the exports to the United States. It is a sign of progress that the government of the province recently distributed more than 1,000,000 grape cuttings, remitting for ten years taxation on new vineyards. A large American commission house has set up a branch in Smyrna, helping to relieve the difficulties of American trade, which may be enumerated as long transit, lack of banking facilities, lack of direct shipping lines, and the unwillingness of American dealers to give credit until goods are received. A Greek company has started a line between New York and the Levant, removing the necessity of transfer at Liverpool, Hamburg, or Naples.

There are three chief ports at the east end of the Mediterranean, — Alexandretta, Beirut, and Haifa. The consul at

Fig. 222. Smyrna and its harbor

Alexandretta reports the establishment of agencies by an American sewing-machine company, and a strong trade. Alexandretta is visited by steamships of nine lines and by many tramp ships. The exports of Beirut are over $6,000,000, including silk to France, barley to the United Kingdom, and wool to the United States (from which the largest import is sewing machines). Safes and roll-top desks sold in Beirut show the modern movement. Here is a center of education due to the work of American missionaries. There is also at Beirut a government industrial college.

In the interior are Antioch, Tarsus, and Aleppo, all bearing ancient names. These are now centers of industry, textiles leading, and flour mills being in operation in all. At Aleppo work in embroideries was started by missionaries among the women and girls, and thousands of persons are now employed. Aleppo is the emporium of northern Syria and receives caravans from Diarbekr, Bagdad, and other places. It is joined to Beirut by

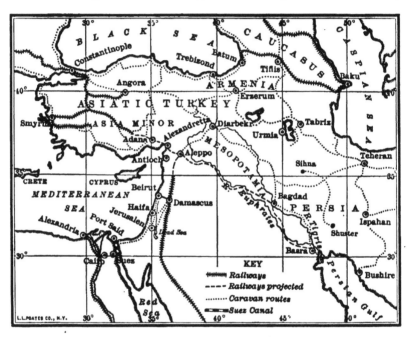

FIG. 223. Routes and commercial centers of southwestern Asia

rail, has 9000 looms, and its exports and imports are not far from $20,000,000 per year. At Urfa is a carpet industry, — a German charitable enterprise. Even in Palestine enterprise has grown up. A new harbor is under way at Haifa, and from this port oranges to the value of several hundred thousand dollars are exported. The making of mother-of-pearl ornaments is an industry of Bethlehem, and tanning is carried on at Hebron. Damascus, with 250,000 people, is on a plain made fertile by the streams of Anti-Lebanon, is connected by rail with Beirut,

and has an electric tramway in operation. Even Medina, far
away in the Arabian peninsula, a center of Mohammedan pil-
grimage, has waterworks and electric lights.

Trebizond is the Black Sea port of Asiatic Turkey. The
trade of its consular district is $28,000,000 per year. Nine
lines of steamships call here, and it is the terminus of the caravan route to Persia, — a country to which belongs about 30 per
cent of the foreign trade of Trebizond.

The Mesopotamian center is Bagdad. A railway will soon
join it to Aleppo on the west, and it has a natural route by the
Persian Gulf to India and the outside world. New steamers
now ply the Tigris and the Euphrates. The foreign trade in
1907, by sea alone, via the port of Bassora, was $15,000,000.
Owing to a short crop in India, Mesopotamian wheat went
thither, and also to Marseilles and London. There is a large
import of cotton goods, the natives sending their own buyers
to Manchester since they know what Arabian and Persian markets require. India is now competing with the United Kingdom in the cotton trade of the Bagdad region. The United
States in 1907 imported goods to the value of $6,500,000 from
Asiatic Turkey.

311. Small commercial centers in Arabia. In the east, at
the entrance to the Persian Gulf, is Oman, whose capital is
Maskat. The United States is one of its best customers for its
largest export, dates. There is a foreign trade of over $3,000,-
000. In the south, at the entrance of the Red Sea, is Aden,
a British possession, a large distributing point, and one of the
chief coaling stations in the world. Farther north on the Red
Sea is Yemen, which in ancient times was called Arabia Felix.
Sana is its interior city and Hodeida and Mokha are its ports,
the latter bestowing its name upon the coffee of the region.

312. Persia. Here is one of the most isolated and backward
countries, even of Asia. Its relations to the British Empire and
to Russia, however, illustrate important principles in the development of commerce. The conceded British sphere of influence
is in the south, the Russian sphere is in the north, and a

neutral zone lies between. The total foreign trade in 1907 was $70,000,000, of which Russia controlled 57 per cent and Great Britain and India more than half the remainder. Russian railways coming to the Persian frontier refuse, with the exception of tea, to carry goods from other countries. Petroleum from the Baku region and beet sugar are large imports from Russia. Persian cotton goes into Russia on one tenth of the usual duty. Russian goods sent to Persia are favored by rebates, and Russian banks in Persia give favorable terms to purchasers of Russian goods.

Southern Persia is served by seven steamship lines, which reach the ports of the Persian Gulf. The chief export from Persia to the United States is in rugs and carpets of native manufacture. The Persian authorities endeavor to check the use of mineral dyes and thus preserve the quality of one of their typical products.

313. India. India is primarily the central peninsula of southern Asia and includes the country lying northward to the Himalaya, northwest to the Sulaiman Mountains, and eastward to the ranges of Burma. Here, however, Burma is included because it is organized as a part of British India. The population is nearly 300,000,000. It is important to study the courses of the Indus, the Ganges, and the Brahmaputra rivers. Along these streams is the great plain of India. North of it are the highest of all mountains, and south of it, filling most of the peninsula, is the plateau known as the Dekkan. On the east the plain extends northeast into Assam, and it forms southward the delta of the Ganges. On the west it reaches north over the plains of the Indus in the country known as the Punjab, and south to the mouth of the Indus.

Except in the Punjab the densely peopled parts of India are south of latitude 30 degrees, and hence are hot and unsuited to colonization and permanent residence by white men. Government officers and missionaries require long furloughs to restore their physical vigor. Here then is a remarkable example of industrial and commercial development, now in progress, for about one fifth of the world's population, under British direction.

FIG. 224. Bombay harbor, with mail boat arriving

This is well shown by the 30,000 miles of railway which India now has, — much more than all other Asiatic countries combined. These roads are specially developed on the northern plains, — from Karachi, an important seaport near the mouth of the Indus, to Lahore, a city of industry and the capital of the Punjab. Thence a branch runs to Peshawar, in the northwest mountain frontier. The main lines pass eastward into the valley of the Ganges and run on either side of the river to the plains of Bengal and to Calcutta. On the west coast a line runs south as far as Bombay. On the east a line reaches from Calcutta, near the shore, almost to the southern tip of the peninsula. The central plateau is less densely peopled, but is crossed by several railways. From Bombay on the west coast a line runs eastward to Calcutta, and others trend southeast to Hyderabad and to Madras. The coast line is little broken, and good harbors and great marine cities are few. The latter are, however, the greatest cities of India, the first three being Calcutta (1,026,000), Bombay (776,000), and Madras (509,000). In 1907–1908 Calcutta was first in foreign trade, Bombay second, Karachi third, Rangoon fourth, and Madras fifth.

Agriculture is the largest occupation, and irrigation is widely practiced. While the country is essentially tropical, the altitudes on the lower mountain slopes and in the Dekkan moderate the heat. Hence much wheat is grown in the Punjab and on the Dekkan plateau. The outlet for the former is Karachi, and for the latter Bombay (sect. 11). The climate and soil of the Dekkan make India a large producer of cotton and afford raw cotton both for export and for several millions of spindles (sect. 23). Bombay is also the port for cotton, and has both cotton and flour mills. Its situation on the west coast gave it great growth on the opening of the Suez Canal, which at once placed it on one of the trunk shipping routes of the world. Rice is the chief crop in the lower lands of the Ganges and Brahmaputra, — the vast and rich plains which form the hinterland of Calcutta. Rice and millet are largely the food of the natives. Rice is an important export, and tea also, whose special region is Assam,

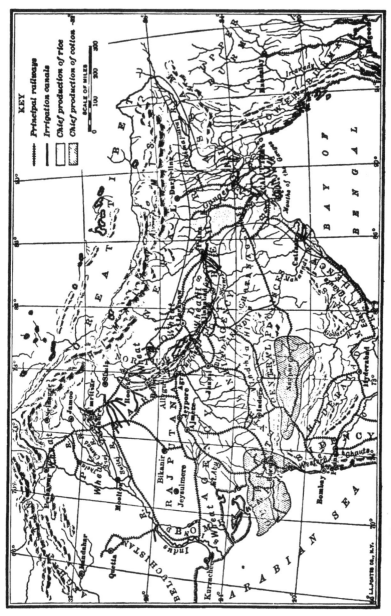

FIG. 225. Chief railways, irrigation canals, and leading agricultural products of India

392

in the region of the Brahmaputra. Jute and silk are also among the textile products and industries.

Burma lies to the east and is a part of the Indo-Chinese peninsula. Its chief city is Rangoon, its artery of traffic is the Irawadi River, and its prime product of the soil is rice, which is largely exported. Closely related to India is Ceylon, whose fields afford tea and whose great mineral product is graphite. India is now probably the most important commercial country

FIG. 226. Harbor of Calcutta

of Asia. China may have greater resources but is more backward, while Japan has much less natural wealth and its native population is far more progressive. As the people of India rise in standards of living they will require more at home, and their export trade may not see great growth.

314. Indo-Chinese peninsula and the East Indian islands.
The Indo-Chinese peninsula includes lower Burma, the French dependencies in this part of Asia (sect. 270), the kingdom of Siam, and the Malay Peninsula. Siam lies between Burma and the French possessions and has Bangkok as its capital. As in Burma, so in Siam, rice and teakwood are the noteworthy exports. The ships that visit Bangkok rank, first, German ; second,

Norwegian; third, British. The Malay Peninsula contains various protected states and British settlements. The most special product of world-wide importance is tin. Typical of a tropical land, though less important, are spices and gums.

The entire Indo-Chinese peninsula has nothing else which compares — in its meaning for commercial geography — with the development and trade of Singapore. This city was founded

FIG. 227. Singapore, Collyer Quay

in 1819 by an Englishman. It is near the south end of the peninsula, and all ships that go east or west around the southeast coast of Asia must pass it. It is a free port, and as a place of distribution (*entrepôt*) and a coaling station is visited by the steamers of more than fifty lines. We may profitably associate Singapore with Hongkong, Aden, Malta, and Gibraltar in their relation to the British Empire.

Of the great islands of the East Indies control is held by the nations of Europe, with the single exception of the Philippines.

These latter islands, having passed from centuries of control by a backward European nation to the keeping of a great commercial power, are now undergoing development in government, in education, and in natural resources. Thus they will represent, as India does to-day, the impress made by a northern race upon a dark people and a tropical land. The largest commercial development among the islands of the entire East Indian group has taken place under Dutch rule, particularly in Java (sect. 283).

315. China. The Chinese Empire is larger than Europe and has an equal or greater population. Its foreign trade is somewhat less than that of Switzerland. Nevertheless China is of great interest to commerce for its present and especially for its future. It has every kind of surface and every grade of density of population. It has great seaports, also interior centers of caravan trade, which are rarely visited by foreigners. China proper is in the southeast and is drained by the Yangtse and Hoangho rivers. Vast plains spread along their lower courses, and about their upper parts are hills and mountains, but nearly everywhere cultivation is intensive and population large.

The winters are cold, especially in the interior, in the higher regions close to the lofty plateau of central Asia; but the summers are warm and moist because the country feels the effect of the monsoon winds. Subtropical products are therefore abundant in the south, while to the north the country runs far into temperate latitudes. In this respect the range resembles that of the United States. In other respects greater contrasts could scarcely be imagined, — dense population, great antiquity, the utmost conservatism, primitive methods of tillage and manufacture, and few railroads.

Spheres of influence and concessions for trade and railway building have gradually been gained by western governments and capitalists, and this vast land is slowly opening to the world. Now the Chinese government maintains some hundreds of students in American and European schools for the purpose of assimilating western knowledge and practice. Silk, tea, cotton, and rice are the typical products of Chinese soil, and all of these except rice are important in the export trade. The mineral resources are known to be enormous, and the deposits of coal are more extended and perhaps offer a larger total of fuel than even the coal fields of the United States. The people are sturdy, industrious, and of large capacity when once awakened

FIG. 228. Eastern Asia

to effective use of their resources. These conditions open high possibilities for Chinese commerce, since a numerous people and a rich country must in time both produce much and require much. The internal commerce is checked by the lack of roads, while the coasting trade is much larger than the foreign commerce.

Shanghai (651,000) is the greatest seaport of China in the volume of its commerce. It is near the mouth of the Yangtse and has a hinterland almost without rival. Peking (700,000) and Tientsin (750,000) may be named together, — the former the capital, the latter its port, with a considerable foreign trade. Canton (900,000) in the south is perhaps the largest Chinese city, but its importance as a port is shared with the neighboring Hongkong, which belongs to Great Britain. The population of Hongkong, mainly Chinese, is over 300,000. Tonnage amounting to more than 11,000,000 enters and clears each year, making this one of the major seaports in the world.

FIG. 229. Harbor of Hongkong

Many other ports and interior cities of great size are little known in America. Of these, two may be named, — Hankau (778,000), a river port of large trade on the Yangtse, 600 miles from the sea, and Chungking (702,000), far up the same river and a center for western Chinese trade. All population figures for Chinese cities are more or less doubtful.

Of four provinces on the north and west, only one, Manchuria, with a temperate climate and wide areas of fertile soil, enters largely into the agriculture and commerce of to-day. The others, Mongolia, Chinese Turkestan, and Tibet, are remote, sparsely populated, with caravan trade and a few ancient centers of life such as Kashgar and Yarkand in Turkestan.

316. Japan. This island kingdom stretches far up and down the Asiatic coast, and was, until about fifty years ago, closed to foreign commerce. Gradually its ports have been opened, its people have awakened, and now, more than any other oriental race, it has entered into modern life and industry. It has been an earnest student and close imitator of western nations, sends its young men to British and American universities, has American and European professors in its imperial university, and teaches its children the English language. Its people have shown themselves diligent, skillful, and courageous, and have quickly brought their country to the rank of the great powers.

The islands are mountainous, the proportion of arable lands is small, and the population dense; but the tillage is intensive, all waste matter is saved, and even the fish of the sea are gathered to give the fields fertility. The chief food products are rice, wheat, barley, and soy beans. Silk, tea, and camphor are typical products, and the chief minerals are iron, coal, and copper. These Japan is now competent to utilize by her own skill, and this is shown in the shipbuilding by which the empire has built up its powerful navy and large merchant marine. Other notable industries are represented by matches, paper, cotton, silk, earthenware, and lacquer ware. The largest trade is with the United States, the United Kingdom, China, Korea, British India, and Germany. Japan has virtual control over Korea and fully shares with Russia influence in Manchuria and eastern Asia in general.

317. Oriental countries and American trade. China is already a considerable buyer of American products. This trade is in its first stages because the tastes and ideals of the people are only beginning to change, and because transportation is poor and the people are shut out from the modern world. Future development is based on resources of soil and mines, which are vast; and on the numerous population, which is by nature strong in body, vigorous in mind, industrious, and given to punctual observance of business obligations. Thus the canceling of contracts by a manufacturing company for reasons deemed sufficient in America is unfavorably regarded in China.

It is natural to ask whether the application of so much latent energy, with the aid of education, machinery, and good transportation, will flood the world with cheap goods and swamp the industries of the white races. It is to be remembered that changes in China will be gradual, giving time for adjustment. Progress in industry also will bring with it higher standards of living, creating a large home market and requiring higher wages for the native worker. It is proper to consider the vast home market of the United States with less than 100,000,000 people, and the future home trade of China with 400,000,000 people.

It has been sought, especially by the United States, to preserve the integrity of China and to give all nations equal commercial rights in that country. The United States has large trade in Manchuria, especially in tobacco, oil, flour, and cotton piece goods, and it is highly important to keep the "open-door" policy, which means that all the chief trading countries are, by treaty with China, entitled to the benefits of the "most-favored-nation" treatment. Following the Russo-Japanese war there appears to have been a strong effort by Japan to gain commercial monopoly in Manchuria. Among the measures used are the promotion of Japanese immigration, the control of the railways and favorable rates for Japanese shippers, the abolition of certain taxes which are still collected on American imports, and free carriage or nominal charges on freight sent by ship.

As a large part of American trade with China has been in Manchuria, these developments are important and have been investigated by agents of the Department of Commerce and Labor. Personal effort has been little used in Manchuria, American cottons selling by reason of their merit. It was reported in 1907 that but two American merchants were residing in Manchuria, a region whose trade with this country runs into the millions. At the same time a large Japanese corporation has many agents. Only one middleman comes between the Japanese producer and the Chinese market, as against six or seven in the case of American goods, and the Japanese in addition gives favorable credit conditions and uses every means of winning the favor of the Chinese merchant.

Japan may be considered as a competing nation, not only in Manchuria but in other parts of the Orient and in all lands. It must be remembered that her natural resources are limited unless she extends control over large continental regions, — a policy which seems to be shown in her efforts in Manchuria and her incorporation of Korea into the empire. She is favored by her inventive capacity and skill in imitation, by energy, and by an almost fanatical patriotism. Japan's governmental organization of industry and commerce is so complete that she has been called a " national development company."

As we send much cotton to European mills, which work it up and win the chief profits, so we are sending raw cotton to Japan, helping that country to supplant, in part, our cotton-goods trade in China, largely by putting goods of inferior grade on the market at lower prices, and by such means of discrimination as have been described. It is stated [1] that American school-books have been printed in Japan, and that German bristles and bone from Chicago stockyards have been made into brushes in Japan, to undersell French goods in the shops of New York. These things do not point to Japanese supremacy, but they show the need of energy and wisdom on the part of the American manufacturer and trader.

318. Asiatic Russia. Siberia stretches from the Pacific Ocean to the Ural Mountains and from the plateau of central Asia to the arctic seas. It has nearly 5,000,000 square miles and about 7,000,000 people, having been gradually acquired by Russia during the past three centuries. The southeast and east are mountainous, while the north and west form one of the few great plains of the world. The north is tundra, the next belt is forest, and in the south are soils and climate which will permit large agriculture of the temperate-latitude type. Vast rivers flow down to the arctic waters, where the frozen seas

---

[1] Bolce, New Internationalism, p. 177.

render them of little commercial value. The great route for commerce is the Trans-Siberian Railway, joining St. Petersburg and all Europe with Vladivostok and the Pacific Ocean. The mineral resources are known to be large. Thus with agriculture,. forests, and minerals this great country awaits only the fuller application of intelligent industry and the achievement of more complete communication.

Russian central Asia is the region inclosed by Siberia and Persia, the Caspian Sea, and Chinese Turkestan. It is a region

Fig. 230. Vladivostok, harbor and terminus of the Siberian Railway

of desert wastes and fertile oases and valleys, of ancient caravan routes and historic marts. Through this region Russia has put a railway, 1871 miles eastward to Merv, Bokhara, and beyond Samarkand ; and from Merv a spur reaches southward toward the Afghan and Persian frontiers. By the trunk line an important export of cotton is brought to the frontier, and the system aids the Russian government in extending its military power and its commerce into central and southern Asia. The Russian province of Caucasus has its chief centers of commercial interest in Tiflis, a trading town of ancient renown, and in Baku, the place of the great petroleum field.

CHAPTER XXV

AUSTRALIA AND NEW ZEALAND

319. General statement. These and certain other lands in the South Pacific are sometimes included under the name Australasia, — a term which means *southern Asia* and expresses close physical relation to the Asiatic continent. Australia and New Zealand are the most important commercial regions in this part of the world, and are of special interest because they supply food and raw materials to older lands, and because they administer government in ways which are considered somewhat socialistic.

320. Physical features of Australia. The length of this land is over 2000 miles and its breadth is above 1000 miles. It is therefore a continent rather than an island. The coast line is uniform and quite unlike that of Europe or North America. Its surface is more even than that of any other continent, its greatest elevation being about one half that of Mont Blanc. The chief mountains border the coast on the east and southeast, and the highest parts are often called the Australian Alps. Westward the surface descends to a vast low plateau, broken by occasional elevations in the interior and near the western coast.

The east is reached by moist winds from a warm sea. Rising upon the cool uplands, they give abundant rains, but are shorn of moisture before reaching the interior, which is extremely dry. The rainfall varies on the east coast from 25 inches at Melbourne to 50 inches at Sydney and 80 inches on the tropical shores of Queensland. Much of the south and west coasts has less than 20 inches and most of the interior less than 10 inches.

Most of the continent has no drainage that reaches the sea. The exception is found in many short rivers on the east slope of the mountains, and in the combined waters of the Darling and Murray rivers, which descend the west slope of the mountains and enter the sea on the south.

402

321. Commercial conditions. Chief among these is isolation. No other advanced country, except New Zealand, is so far from the main routes of travel and the great centers of commerce. The distances are so great that without modern methods of transportation and communication Australia could have little to do with the rest of the world. Either Canada or Argentina is much nearer to London, Hamburg, or Genoa than is Australia.

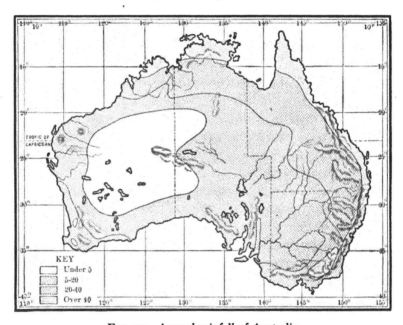

FIG. 231. Annual rainfall of Australia

Australia, like Canada and the United States, had but a small native population, and thus contrasts with India, Java, South Africa, or Brazil, offering a fresh field for European influence and colonists. Australia has a more homogeneous population than a land which, like the United States, has been in the track of emigration from the crowded nations of Europe. Her climate is tropical and warm-temperate, a condition favorable to agriculture and life in the open air. A peculiar condition is the vast proportion of arid land in the interior, leaving a habitable fringe along the sea and leading to a concentration in seaboard cities which is remarkable when the small population and continental extent of the country are considered.

In government Australia may be compared to Canada. The six states, Victoria, New South Wales, Queensland, South Australia and West Australia, and the neighboring island of Tasmania, have been united since the first day of the century, January 1, 1901, into the commonwealth of Australia, under the British crown. This union introduced free trade among the states and unity of administration in foreign commerce.

322. Government activity and social conditions. The railways are chiefly built and administered by the several states, and the telegraph is conducted by the commonwealth as a part of the postal system. Public ownership was a necessity if the country was to have modern methods and not be retarded in development. Railways and telegraphs must run through uninhabited regions where private capital could get no returns. These utilities were financed by British capital. But the large public debt thus incurred has behind it productive utilities and is not a useless burden, as are many national debts resulting from military expenditure.

The paternal functions undertaken by the government make necessary a vast number of civil servants and cause at the same time widespread political interests among the people. Thus the labor movement is claimed to have "given Australia its distinctive place among the nations." The labor party has naturally had much to do with the conditions of industry and commerce. It stands for such provisions as the eight-hour day, arbitration courts, indemnity for injuries, and for old-age and invalid pensions. Workingmen are independent and have full opportunity to rise in means and influence, in a land in which distinctions of money and class do not greatly count.

Several states had adopted the policy of pensions for old age, and in 1909 this became the law of the commonwealth, with the qualifying age of sixty-five years and twenty years of residence. The commercial policy is protective, the average duties being about 15 per cent. The native producer is in some cases further protected by high rates of freight. The tariff is planned to prevent monopolistic control so far as possible, to prevent dumping by foreign trusts, and to protect the workingman. In this aspect the name " the new protection " is given to the present tariff system. A working feature is the laying of an excise on home manufactures equal to one half the duty on similar imports, the excise being remitted where conditions of labor are decided to be fair and reasonable.

Another important factor in the commercial life is the determined policy of a " white Australia," due to the fears of depreciated labor and inferior society, should this open and attractive land be overrun by Asiatics, who live within a few days' sail. To this end the Kanaka laborers, imported to till the sugar plantations of Queensland, have all been deported, the industry being sustained during the change by bounties given on sugar grown by white labor. The industry still thrives, for as the cost of labor rises, improved methods of cultivation are adopted. Even though the tropical belt be developed more slowly, Australians are agreed in holding the country for the higher standards of the white race, seeing the opportunity to build a truly British nation in the southern hemisphere. It is sought to retain strong bonds with the empire by giving a preferential tariff on all British goods, and the problem of national defense is regarded as vital to future safety.

323. Transportation and communication. Australia of necessity reaches the rest of the world by water, and is yet peculiarly dependent on ships for transit between her states. Eastern New South Wales, Victoria, and the adjacent part of South Australia offer the only network of railways in the commonwealth, and the chief cities are all by the sea. Unfortunately, also, the railways of the various states are not of uniform gauge. Short lines run inland as follows : from Port Darwin to Pine Creek in the north ; from Fremantle and Perth to Kalgoorlie in the west ; from Port Augusta to Oodnadatta in the south ; and from Sydney, Brisbane, and other ports in the east. These inland spurs have been built for mining purposes, and in some

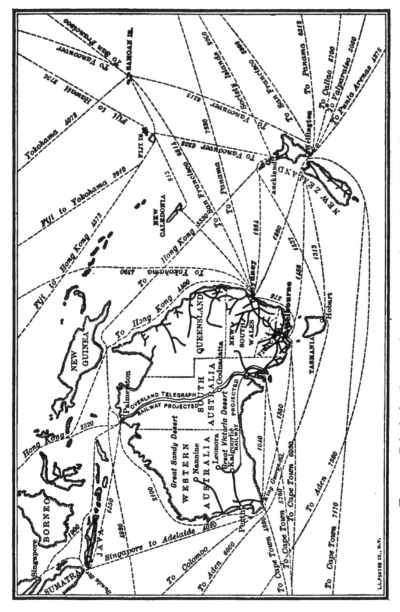

FIG. 232. Principal railways and ocean routes of Australia and New Zealand

406

cases to move the flocks to unpastured ranges and save them from starvation in time of drought.

It is proposed to build three transcontinental lines : (1) from Port Augusta in South Australia westward to Kalgoorlie, Perth, and Fremantle ; (2) from Oodnadatta northward to Pine Creek and Port Darwin ; (3) from Bourke in New South Wales northwest to Pine Creek and Port Darwin.

The union of telegraph stations with post offices involves economy of management, and notwithstanding long distances and small population the rates are much cheaper than in the United States, and each inhabitant averages two to four telegrams per year, as compared with one message in the United States. Ocean cables join Australia to other lands by way of Java, and a cable also extends from Vancouver to Queensland. Need of more ample cable service is felt, the cost now being so high that foreign news is much condensed, thereby giving a limited knowledge of markets and the affairs of the empire.

Ocean steamship lines reach all Australian ports, joining them with Europe, Asia, America, South Africa, and New Zealand. The great highway to Europe is through the Suez Canal. Australia will feel the shifting of the streams of the world's commerce on the opening of the Panama Canal, a route which will greatly shorten the voyage from London to Melbourne, Sydney, or Brisbane.

324. Grazing industries. Like other new and sparsely settled lands, such as western America and Argentina, Australia has had large interests in sheep and cattle, and still keeps her rank as the first wool-growing country in the world. Australians believe that their pastures are specially favorable to the quality of wool and that their supremacy will be permanent. In the early days of the colony pure merino sheep were, with much difficulty, obtained from Spain, and the foundation of future prosperity was laid.

Wool growing has largely influenced the life of Australia, not only in the solitary life of the ranges, but in the building of railroads and in the concentration of people in the seaports, whence the wool product to the extent of 98½ per cent is shipped to foreign lands. Australia had nearly 88,000,000 sheep at the end of 1907.

At the same time there were over 10,000,000 cattle. Of the butter and cheese, the latter is used at home and the former is largely sent to the United Kingdom. Here, as in other things, there is much state aid and inspection. The government offers instruction in methods, imports blooded stock, and provides refrigerating chambers in ships in connection with its mail contracts.

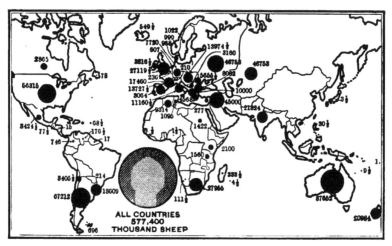

FIG. 233. Number of sheep, in thousands, figures being given for all countries from which dates are obtainable

325. Agriculture. Among direct products of the soil wheat is far in the lead, occupying one half of the cultivated lands. Though the commonwealth may not produce more than Kansas or Minnesota or Manitoba, its export of wheat is large, owing to the small population at home. Oats is the next crop, but is far less important than wheat. Maize is extensively raised, and shows again how far this North American cereal has migrated in modern times. Vine growing and wine making have grown to considerable proportions. Sugar is the chief product of northern or tropical plantations, but there seems to be no reason why cotton, rice, and tropical fruits may not become large industries.

326. Mineral industries. The total mineral product in 1907 was about $140,000,000. Gold is the largest item, and in 1901 the commonwealth produced more than one fourth of the world's supply. Soon after that the relative rank fell because of the great output of the Transvaal. Silver, copper, and tin are important, and there are minor metals and gems, with vast fields remaining for exploration and probable large discoveries.

FIG. 234. Harbor of Sydney

Coal and iron, the minerals which are fundamental to great industrial development, are believed to be widespread and important, but are imperfectly known and slightly developed, with the single exception of coal in the state of New South Wales.

327. Cities and manufacturing. In 1907 the population of Sydney was estimated at 577,000 ; and Melbourne is nearly as large as Sydney. The advanced character of these great centers is noted by all writers, and the two contain more than one fourth of the people of the commonwealth. In 1907 Adelaide, the capital of South Australia, had 178,000 ; and Brisbane, the capital of Queensland, about 135,000. When it is remembered that the population of at least seven other towns ranges from 25,000 to above 50,000, the concentration in cities is seen to be remarkable for a new country of small population.

Manufacturing is in its early stages and has been compared to that of Canada fifty years ago or to that of Argentina to-day. Growth in this department is now fostered by the federal policy of protection. The simple preparation of primary products holds first place, as in other new lands; examples are soap, brick, brewing, and wool scouring. There is progress and even some foreign trade in agricultural and mining machinery.

328. Trade. There has been great expansion of internal commerce, owing to the removal of tariff barriers by the federation of the several provinces. The total over-sea trade of Australia is not far from $150 per capita of the population, a remarkable showing for the productiveness of the country and the enterprise of the people. An English writer has not hesitated to call Australia " by far the richest of all the colonies." Much the largest export is wool, valued in 1907 above $140,000,000. Gold specie and bullion are second, followed by wheat, copper, butter, skins and hides, mutton, tallow, and various metals. Much the greater trade is with the United Kingdom. Otherwise Australia imports chiefly from the United States and Germany, and exports to France, Belgium, Germany, and the United States. These European countries need her food and raw materials, and she offers to manufacturers in the United States a market for mining and farm machinery and for such American specialties as typewriters, phonographs, and sewing machines. Among the less elaborated products drawn from the United States are timber, tobacco, wheat, and petroleum. It is believed that the new preferential tariff has arrested a tendency to decline in the trade of the commonwealth with other parts of the empire.

329. New Zealand. The islands bearing this name ceased to be a crown colony and became the Dominion of New Zealand in 1907. There are two chief islands, having an area of a little more than 100,000 square miles and a population slightly above 1,000,000, mostly European. The surface of the land consists of lofty mountains, wooded slopes, and fertile plains. Agriculture, grazing, and timber cutting prevail, and meat freezing, tanning, sawing, and butter and cheese making are the principal processes of manufacture. Wool, frozen meat, butter and cheese, and gold are the chief exports. After the United Kingdom and Australia the United States has the most important place in the foreign trade of the dominion. New Zealand is seeking to solve the problem of popular government through the same types of progressive legislation which prevail in Australia.

CHAPTER XXVI

AFRICA

330. Conditions affecting production and trade. These may be analyzed as follows :

1. The close relation to Asia and Europe, Africa being joined to the former by an isthmus and separated from the latter by a narrow sea. The ancient life and commerce of the Mediterranean was shared by North Africa, but did not reach far into the continent, perhaps on account of the Sahara. With such a history we may contrast the growth of great nations in a few centuries in America.

2. Configuration. Africa has almost unbroken shore lines and few harbors. Its coastal lowlands are narrow and its broad interior is a plateau. The rivers go down by rapids and falls and soon reach the sea, putting a great check on navigation between the coast and the interior. There are few mountain ranges of great length or height, to stop the winds, condense the moisture, and diversify the surface and climate.

3. Climate. Much of Africa is too dry for a good cover of forest or crops, and much is absolute desert. It is the most thoroughly tropical of all the continents, South Africa and some of the higher lands alone being temperate. The prevalence of tropical diseases is the greatest obstacle to the industrial and commercial progress of central Africa.

4. The character of the native people. Fanaticism and intolerance have prevailed in the north, and dark and ignorant savagery in the center and south, until far into the nineteenth century. Such conditions pass slowly ; and without intelligence, industry, and security for property and life, production is checked and commerce is dangerous and profitless.

5. The overcrowded commercial nations of Europe have sought outlets here, seeking to widen their trade, to increase

national prestige, and, in South Africa at least, to provide homes for surplus population. This is the force which has given to Africa the beginnings of commercial life.

331. European dependencies in Africa. The political map of Africa is a patchwork. Regions are here reviewed by nations.

1. Great Britain. Egypt has been subject to Turkey, but is now under British policy, as is the eastern Sudan. On the Indian Ocean is British East Africa, whose capital is Mombasa. These lands give practical British control of the Nile. In West Africa are Gambia, Sierra Leone, Gold Coast, and the highly important Nigeria, occupying the lower Niger basin. British South Africa extends far north between German and Portuguese lands to the sources of the Kongo and to Lake Tanganyika. With a single break in German East Africa, the British can finish the Cape-to-Cairo Railway in lands which they control.

2. France has Algeria, including Tunis, the western and central Sahara and the Libyan Desert to the border of the Nile basin, the central and west Sudan, and the French Kongo. By active exploration and improvements the French in the last fifty years have mastered most of North Africa. They also hold a small piece of the Somali coast at the opening of the Red Sea, and the great island of Madagascar. In square miles, France is at the front in Africa, but British lands are superior in value.

3. The German Empire has four dependencies: Togo, on the Gulf of Guinea; Kamerun, a rich province between Nigeria and the French Kongo; German Southwest Africa; and German East Africa. Southwest Africa is largely desert, because the winds from the Indian Ocean have parted with their moisture in passing the eastern highlands. East Africa is of greater value, reaching from Lakes Victoria Nyanza, Tanganyika, and Nyassa to the coast opposite the British island of Zanzibar.

4. Italy controls the Somali coast for several hundred miles south of Cape Guardafui, and the province of Eritrea on the southwest shore, after passing within the Red Sea.

5. Portugal has three areas: Portuguese Guinea on the west coast; Angola on the west coast south of the Belgian Kongo;

and Portuguese East Africa, with a long coast line and including an important region, the lower basin of the Zambesi.

6. Other dependencies and nations. Turkey retains control of Tripoli, Belgium has the greater part of the Kongo basin, Spain has a region adjoining Morocco on the Atlantic, and a small tract on the Gulf of Guinea between Kamerun and the French Kongo. There are three independent nations, all of small importance at present : they are Morocco, Abyssinia, and the republic of Liberia.

This definite partition of Africa dates from the Berlin conference of 1885 and subsequent treaties between various European nations. Asia has a few great sovereignties, — either native, as China and Japan ; or colonial, as India and Russia in Asia. South America, with slight exception, is composed of independent republics. Australia belongs solidly to the British empire. Africa is the one continent which is minutely divided among European powers, which thus take responsibility for its development in civilization and commerce ; Great Britain, France, and Germany are accomplishing the most for the progress of the " dark continent."

332. Africa viewed in belts of latitude. The continent can be much better understood if analyzed in broad regions than if studied in its broken political divisions. Here it is natural to take, —

1. The Mediterranean border, including Egypt, Tripoli, Algeria, and Morocco, under British, Turkish, French, and independent rule. Tripoli is little more than a desert, but the city is the terminus of a route of caravan trade across the Sahara. Morocco will have commercial importance when misgovernment and semi-savagery give way to security, when its natural wealth in minerals and soils can be developed by modern processes and its products carried to market by other than animal portage.

Egypt, the seat of one of the oldest civilizations, is one of the chief commercial countries of Africa, and is of growing importance. Nominally it has 400,000 square miles, but its population of more than 11,000,000 really occupies a cultivated and

iding

orto
basi
and a
d the
all e
i, and

con
Euro
either
Russa
sed of
British
ivided
for its
rance,
ess of

t can
than
tural

poli,
and
the
ara.
ient
alth
ses
ge.
of
m-
ou-
d

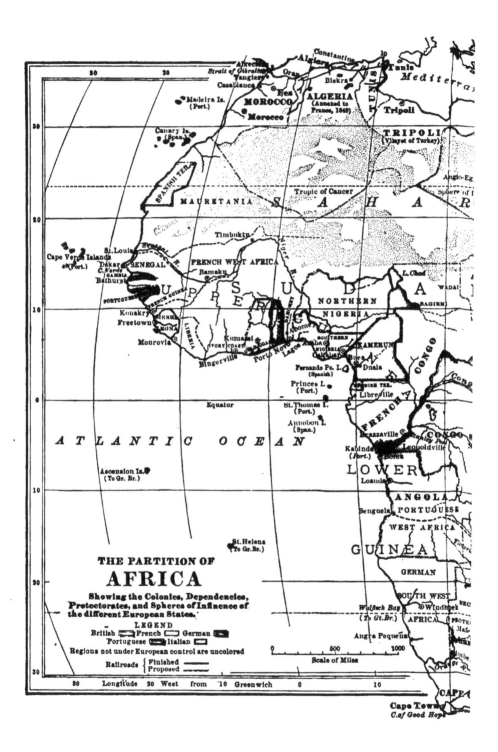

THE PARTITION OF

AFRICA

Showing the Colonies, Dependencies,
Protectorates, and Spheres of Influence of
the different European States.

LEGEND

British | French | German
Portuguese | Italian

Regions not under European control are uncolored

Railroads { Finished ——— Proposed ———

Scale of Miles
0   500   1000

Longitude 20 West from 10 Greenwich 0 10

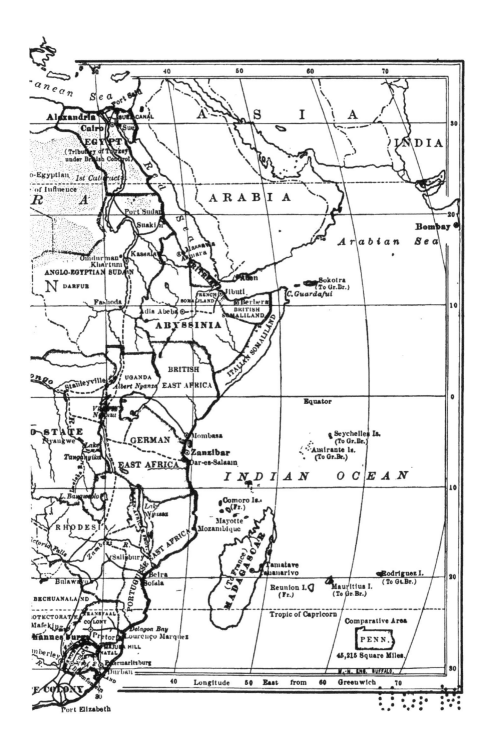

settled area of about 12,000 square miles, which makes the density much greater than that of even Belgium or England. Nearly two thirds of the people engage in agriculture. The winter crops are cereals and the summer crops are cotton, rice, and sugar. By controlling the river, and particularly by the

Photograph, copyright 1903, Wm. H. Rau, Philadelphia

FIG. 235. Nile dam at Assuan

great dam at Assuan, waters are stored and perennial irrigation is extended over large areas. This will increase the products, and in consequence there will be greater population, more workers, and more exports. Cairo is the capital and the chief city not only of Egypt but of Africa (654,000 in 1907), and Alexandria is the chief port, with over 300,000 people. Over 2000 steamships, with tonnage of 3,500,000, arrive and clear each year at Alexandria, marking the growth of Mediterranean trade.

The Suez Canal, dug by De Lesseps and purchased in 1875 by Great Britain, was opened in 1869. Including 21 miles of lake navigation, this waterway from the Mediterranean to the

FIG. 236. Port Said and the entrance to the Suez Canal

Red Sea is 87 miles long and is traversed annually by over 4000 vessels of tonnage of over 20,000,000. Nearly two thirds of both vessels and tonnage are British.

2. The Sahara, the second latitude belt, extends from the Atlantic Ocean to the Red Sea, and from the Mediterranean countries to the Sudan, or from 18 to 30 degrees north latitude. In parts it is a waste of sand, in parts a desert of rocks, and contains 3,500,000 square miles, approaching Europe in size. It may be considered as continuing through Arabia, Persia, and Turkestan to Mongolia, thus almost crossing the greatest land mass of the globe. The desert reaches the Atlantic because the winds are offshore rather than on-shore. They are

the northeast trade winds crossing the Mediterranean to Africa, but they gather little moisture from this narrow sea, and are moving into a warmer region and hence have greater capacity to hold their load of moisture. The commercial value of the Sahara is in its oases, its routes of caravan trade, the possibility of irrigation where water can be found, and the discovery of useful minerals.

3. Sudan. This third belt is south of the Sahara, and is itself bordered on the south by the Gulf of Guinea and the northern rim of the Kongo basin. Its latitude is from 4 to 18 degrees north. It contains many political divisions but belongs largely to the French and the English. The term is sometimes used to include Abyssinia and thus is carried from the Atlantic Ocean to the Red Sea. From the desert conditions of the Sahara there is a gradual increase southward of rainfall and vegetation, and the capacity for agriculture and ultimate trade is large. In Nigeria and other parts bordering the Gulf of Guinea rainfall is heavy and conditions of tropical luxuriance prevail. The problem is to train the laborer and to secure conditions of safe living for the Europeans, who, for a long time to come, must direct in government and industry. In the central Sudan is Lake Chad, which receives the drainage of a closed basin and thus shows the insufficient rainfall of that part of Sudan. Anglo-Egyptian Sudan covers nearly 1,000,000 square miles. It has much land along the Blue Nile that might bear cotton and wheat, and there are valuable forests on both the Blue and the White Nile. These contain rubber, bamboo, ebony, and fiber and tanning materials. Khartum is the capital, and here is Gordon College, named for the British general who reduced the region to order; it is significant for commercial and social growth that this school has in its shops nearly 200 boys under training in industrial work. A railway has now been extended to Khartum from Cairo.

4. The heart of equatorial Africa is the Kongo basin, which is larger than that of the Mississippi. It has a hot climate, is well watered, and is densely forested. It is largely occupied by

the so-called Kongo Free State, which, however, was placed under the direction of the late Belgian king, whose administration aroused serious criticism. In 1907 the state was annexed by Belgium, an act which in 1909 had been recognized only by Germany. This division has over 900,000 square miles. It produces rubber, palm nuts, palm oil, and white copal; and tobacco, cocoa, and coffee are readily grown. The total foreign trade is about $20,000,000, rubber being much the largest item. Cottons form the chief import.

The Kongo is navigable 100 miles from the sea. Then follow rapids, 200 miles to Leopoldville, and above are 1200 miles of navigable water to Stanley Falls, besides much navigation on tributaries. A railway extends 250 miles around the rapids, and 900 miles now building will join the Kongo River to Lake Tanganyika and Lake Albert Nyanza. These lines will connect the Kongo country with the Cape-to-Cairo Railway.

5. South Africa. This region may be said to begin with the Kongo-Zambesi divide. Portugal, Germany, and Great Britain are the ruling powers. Except on the lower Zambesi the coastal plain is narrow and the interior mainly an elevated plateau, arid in the central and western parts. The British regions of Cape Colony, Natal, Orange River Colony, and the Transvaal, and Portuguese East Africa are the most productive parts; British South Africa, in its eastern half, has the most extensive network of railways on the continent, the nearest rival in development being under French direction in northern Algeria.

333. Agricultural resources of Africa. Dry as much of Africa is, the continent is so large and the climate so warm that the capacity for crops is vast. The problems of development have already been stated, namely, to discipline the native to labor and industry; to acclimate the European who must be the organizing head; to evolve stable conditions of government and security of life and property; and to develop transportation. Temperate-latitude products, as wheat, mingle with tropical harvests in North Africa; every conceivable tropical growth flourishes or may be developed in central Africa, while wheat and other temperate-latitude crops are grown in South Africa. Already a considerable export of maize from Natal is sold in London. More than 1,500,000 pounds of tea were grown in South Africa in a recent year. Here animal industry is emphasized, as in the production both of wool and of ostrich feathers.

FIG. 237. Sorting diamond gravels, Kimberley

334. Mineral resources. General exploration tells much about soils and capacity for crops, but close search is often necessary to find the useful minerals. Hence the mineral wealth of Africa is little known.

The diamonds and gold of South Africa form the best known and most profitable mineral industries of the continent. Diamonds are found along the Vaal and Orange rivers. They were discovered in 1867 through a stone picked up by the child of a Boer farmer. They occur in river deposits and in what are known as "dry diggings" in the "plugs" or filled pipes of ancient volcanoes. The chief place in the diamond industry is Kimberley, in Cape Colony. South Africa now furnishes almost the entire supply of these gems for the markets of the world. They are cut by the craftsmen of Belgium and Holland, and the greatest ultimate market is in the United States. The company controlling the mines limits the output in order to maintain prices.

British South Africa is one of the chief gold-producing regions. The Transvaal in 1907 exported gold to the value of $136,000,000. Johannesburg in the Transvaal is the center of the gold interests and is the largest city of South Africa. In various parts of British South and Central Africa are also found important deposits of iron, copper, and coal, affording, with gold, gems, agriculture, grazing, and a temperate climate, the basis of commercial development.

The British dependency, Gold Coast, on the Gulf of Guinea, is also productive in gold. Algeria has mines of importance for coal, iron, zinc, lead, copper, mercury, and silver. Gold mining is a considerable industry in Madagascar, and other metals are found there.

335. Transportation and centers of trade. The growing framework of transportation may best be understood by reference to the Cape-to-Cairo Railway, which is now completed from Cape Town to Bulawayo in Rhodesia, and from Cairo to Khartum. The intermediate section of about 2500 miles will follow the Nile and the west shore of Victoria Nyanza, and pass, on the east, the south end of Lake Tanganyika. Between these lakes it will traverse German territory. From Bulawayo a line is now continuous to Beira, a port in Portuguese East Africa. Northward from Bulawayo several hundred miles of new road reach the Zambesi at Victoria Falls, cross it by a great bridge, and reach mining territory nearly 400 miles north of the river. A road is projected in German East Africa from the Indian Ocean to the great trunk line. In British East Africa the Mombasa-Victoria Railway is 584 miles long and joins the port of Mombasa with Victoria Nyanza. Projected Kongo connections with the trunk line have already been noticed. On the Nile delta railways branch to Alexandria, Rosetta, Damietta, and Port Said.

In Algeria the main line extends from Tunis westward at some distance from the Mediterranean and sends off many short spurs north and south. From Algiers a road is projected southward across the Sahara to the northern bend of the Niger River. A short line runs in from the coast in Portuguese West Africa, and similar roads are projected from the Gulf of Aden, in Gold Coast, and in German Southwest Africa. In Senegal a line is proposed from the Atlantic coast to the upper Niger. Thus the Cape-to-Cairo line, the lines that join it to the east and west coasts now or in the future, with the four master rivers, will in due time open to profitable commerce every important part of a continent so recently not only unused but unknown.

Algeria, Tunis, Tripoli, and the ports of Egypt are the chief centers of trade on the Mediterranean. On the east coast are Mombasa and the British city of Zanzibar, the latter on an island off the coast of German East Africa. Far south on Delagoa Bay, Lourenço Marquez, a Portuguese city, is said to have the best harbor on the coast of Africa. It is joined by rail to Pretoria in the Transvaal. Continuing south, we find the British ports Durban, East London, Port Elizabeth, and finally Cape Town, all having railway connection with the interior. The ports of West Africa are as yet of minor commercial importance. Timbuktu is a small but ancient city near the Niger in its northern bend, and is the center of the caravan trade of the western Sahara.

CHAPTER XXVII

LATIN AMERICA

336. General view. The studies in this book, after dealing with the home country somewhat fully, passed to Canada, with which relations are close; then to the countries of Europe, the most important commercial region; then to the other continents of the Old World; and now they return to that part of the New World which lies south of the United States. Here are included South America, the West Indian islands, Central America, and Mexico. The last country, like Canada, adjoins the United States and is united to it by close ties of trade.

These lands were mainly discovered and colonized by Spain, Portugal, and France, and the languages of these countries still prevail. As these nations are in part of Latin origin, the term *Latin America* is a suitable general name. Most of the American nations concerned have, during the past century, thrown off their allegiance to European powers and have formed republican governments. The ties of language, of kinship, and of prolonged commercial exchange are, however, strong, and in many respects South and Central American peoples are more closely related to Europe than to the United States.

Recent events indicate growing bonds among all nations in the western hemisphere. Through the Spanish War of 1898 the United States gained new influence in tropical America. The construction of the Panama Canal gives this country a new foothold in the south, not only opening a ready channel of commerce with the Andean countries, but setting an example of skill in mechanical operations, and of thorough sanitation, the influence of which is already fruitful among the Latin-American peoples.

South America, like North America, has its principal slopes and its great commercial rivers on the Atlantic side. This

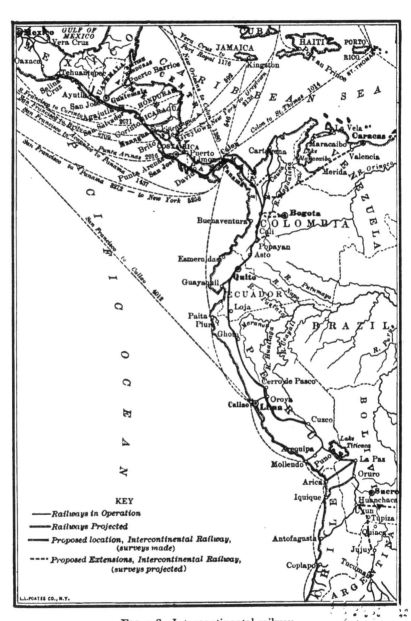

FIG. 238. Intercontinental railway

condition favors commerce with Europe. South America is chiefly
east of the longitude of New York, a condition which also helps
its relation to Europe. Added to this is the fact, unfortunate
for American trade, that the merchant marine, in foreign commerce, is chiefly in the hands of European nations. Thus even

FIG. 239. Building of the Pan-American Union, Washington, D. C.

Italy has a worthy share in the carrying trade to South America,
stimulated by the presence, particularly in the Argentine Republic,
of many Italian settlers and laborers.

337. Brazil. Brazil is the largest and the most populous of
the Latin republics. With 20,000,000 people it has about two
fifths of the population of South America. In its foreign trade
it is, however, exceeded by the Argentine Republic. Brazil is
an agricultural country, and coffee provides much the largest of
all field products, the export amounting to $115,000,000 in
a recent year. The great field is in the southern provinces,
where four fifths of the world's crop is produced. Europe takes

FIG. 240. Average annual production of coffee in thousands of pounds, 1904–1905 to 1908–1909 inclusive

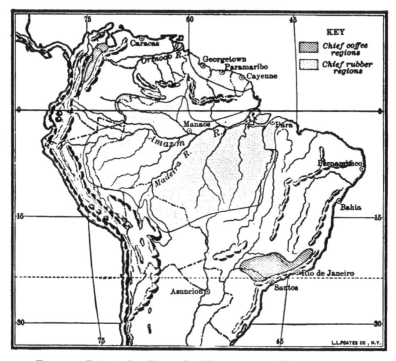

FIG. 241. Principal coffee and rubber regions of South America

57 per cent of the coffee, though some of this is reëxported. The United States takes the remainder, and of this, two thirds goes to the port of New York. The state of São Paulo, the city of the same name, and its port, Santos, form the special centers of coffee production and shipment.

Rubber is the second item of the foreign trade, and is mainly produced in the tropical forests of the Amazon region. Cocoa

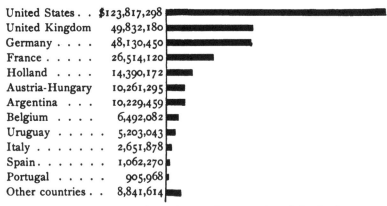

United States . .	$123,817,298
United Kingdom	49,832,180
Germany	48,130,450
France	26,514,120
Holland	14,390,172
Austria-Hungary	10,261,295
Argentina . . .	10,229,459
Belgium	6,492,082
Uruguay	5,203,043
Italy	2,651,878
Spain	1,062,270
Portugal	905,968
Other countries . .	8,841,614

FIG. 242. Exports of Brazil, 1909, by countries of destination

and brazilwood also have a considerable place. The total foreign trade is about $400,000,000, which again shows the undeveloped state of a country having about the size of the United States, with trade less than that of some minor European countries. Development has public aid, however, bounties being offered on wheat, and duties being remitted on certain importations of agricultural machinery, fertilizers, and insect destroyers.

United Kingdom	$48,241,287
Germany	28,007,001
United States . .	22,265,534
France	18,610,398
Argentina . . .	17,922,587
Portugal	9,994,615
Belgium	7,280,007
Uruguay	6,294,057
Italy	5,236,557
Austria-Hungary	2,365,825
Newfoundland .	2,008,641
Switzerland . . .	1,963,169
Holland	1,748,977
Spain	1,522,001
Norway	1,504,933
Canada	953,523
Other countries .	3,771,013

FIG. 243. Imports of Brazil, 1909, by countries of origin

Manufacturing is only in its beginnings, the cotton industry being most important. There are 12,000 miles of railway, and a new road is now

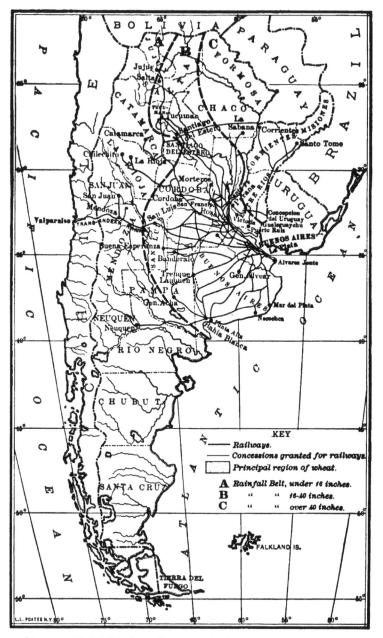

FIG. 244. Rainfall belts, railways, and wheat region of Argentina

426

building around the falls of the Madeira, opening a trade route from Bolivia to the Atlantic, by the Madeira and Amazon rivers. Manaos is a considerable port on the Amazon, 1000 miles from its mouth, and modern ships ply between it and Rio de Janeiro. There are 10,000 miles of river navigation for ocean vessels, and 20,000 miles additional, suited to light craft. A German company has a concession for a cable from Europe by way of Teneriffe. The United States imports heavily from Brazil, but makes small returns. Great Britain and other European nations take far less and yet supply the bulk of Brazilian imports. This is attributed partly to the poorer shipping communications with the United States and partly to American ignorance of the Portuguese language and the historic kinship of Brazilians with the people of Europe.

Rio de Janeiro is the capital and commercial center, being the second city in South America. It is 23 days from New York, while swift and modern vessels reach England in from 14 to 16 days.

338. River Plata countries. The chief of these is the Argentine Republic, third in size of American republics and having a population of a little over 6,000,000. It embraces 34 degrees of latitude, reaching into the torrid zone on the north, and including, on the south, the eastern part of Tierra del Fuego. It resembles the United States in affording products both of temperate and tropical climates. In surface also it may be likened to the prairies and plains of the Mississippi Valley. These stretch northward along the River Plata and its great tributaries and southward through Patagonia, and extend out from the eastern slopes of the Andes.

The centenary of the country's independence was celebrated in 1910, and was marked by an exposition of national industries, especially of methods of transportation, thus touching a fundamental condition of commerce. The republic is almost wholly agricultural. As both coal and water power appear to be small in amount, this condition will probably be permanent. The flatness of the country makes the rivers navigable and favors the building of railways. Barely one tenth of the arable land has yet been used. Even with present development the annual foreign trade is more than $600 000,000, thus exceeding that of Brazil with its greater size and its tropical luxuriance. Argentina took the first rank in export of corn from the United States in 1908, although still the third country in the production of that cereal. Owing to small population, it has a greater surplus, as Canada has of wheat. It is fifth in wheat growing, second in shipments of wool, and first in exports of frozen

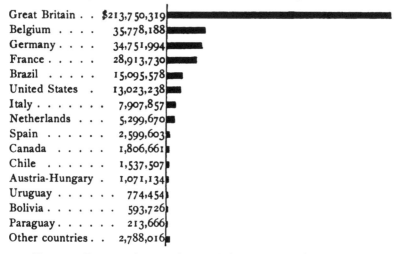

Great Britain . .	$213,750,319
Belgium	35,778,188
Germany	34,751,994
France	28,913,730
Brazil	15,095,578
United States .	13,023,238
Italy	7,907,857
Netherlands . . .	5,299,670
Spain	2,599,603
Canada	1,806,661
Chile	1,537,507
Austria-Hungary .	1,071,134
Uruguay	774,454
Bolivia	593,726
Paraguay	213,666
Other countries . .	2,788,016

FIG. 245. Exports of Argentina, 1908, by countries of destination

meat. These facts show that animal industry and temperate-latitude cereals have the chief place. The area under cultivation increased over 200 per cent between 1895 and 1908, and large immigration and improvement in methods indicate continued growth, as in Canada, whose productive areas and population are similar in extent and numbers. A recent annual output of wine to the value of $12,000,000 points to diversified agriculture, as does the presence of more than 11,000,000 mulberry trees for silk culture. The animal industry is moving west and south toward the drier plains, while tillage is claiming the provinces nearer the river Plata. The forests are chiefly in the warmer regions of the north.

The railways show a mileage of over 15,000. There is a network along the Plata, from Buenos Aires to Rosario, Santa Fe, and northward. From Rosario a road runs far north to Jujuy, connecting with a road now building to the capital of Bolivia. This line opens the grain country and also reaches the center of a considerable sugar industry. Westward a line extends to Mendoza, and now connects, across the Andes, with Valparaiso and the Pacific coast. An important group of roads occupies the plains southward, from Buenos Aires to Bahía Blanca.

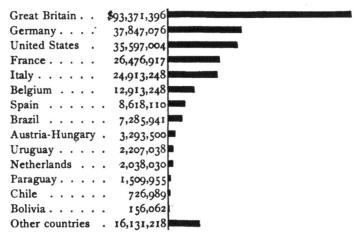

Great Britain . .	$93,371,396
Germany . . . . ʹ	37,847,076
United States .	35,597,004
France	26,476,917
Italy	24,913,248
Belgium	12,913,248
Spain	8,618,110
Brazil	7,285,941
Austria-Hungary .	3,293,500
Uruguay	2,207,038
Netherlands . . .	2,038,030
Paraguay	1,509,955
Chile	726,989
Bolivia	156,062
Other countries .	16,131,218

FIG. 246. Imports of Argentina, 1908, by countries of origin

Buenos Aires, with 1,200,000 people, is not only the chief port, but is the largest city in the southern hemisphere. Five lines ply between Buenos Aires and the United States, but faster ships, and in greater number, carry on traffic with Liverpool, Bremen, Hamburg, Bordeaux, Barcelona, Marseilles, and Genoa. Ocean connection with Italy has grown rapidly because of large Italian immigration and the resulting desire for Italian merchandise. Much of the Atlantic voyaging is triangular. This is true both of Argentina and Brazil. Buenos Aires, Santos, or Rio Janeiro may be the southern terminus, New York the western, while the eastern terminus is some port of western Europe.

With its ample territory, progressive spirit, and European steamship connections, Argentina is an important rival of the United States in the agricultural products of temperate latitudes and a strong competitor in the world's markets.

Uruguay and Paraguay are the remaining Plata countries. They are far smaller, are rich in commercial possibilities, and at the present time but little developed.

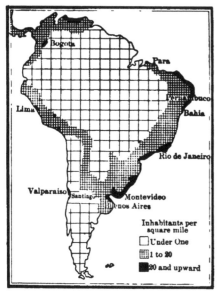

FIG. 247. Distribution of population in South America

339. Countries of the west coast. These are Chile, Peru, Ecuador, and Colombia. The last borders also the Caribbean Sea, and since its greater foreign trade is there, it will be considered with the northern coast. Chile is chiefly on the narrow west slope of the Andes, while Peru and Ecuador hold important territory in the upper Amazon basin. In all these countries both agricultural and mineral resources are large. In South America, as in North America, the western mountains are productive in minerals. Peru has, among other minerals, silver and copper, while Chile is rich in nitrates and copper. Chilean nitrates are used in many lands as fertilizers, the annual output amounting to about $80,000,000. The west coast needs iron and steel, and also textiles and the food products of temperate latitudes. For the tropical growths of this belt there is a market in the United States, and for this trade the Panama Canal will open the way. Already there is a considerable exchange, and American trade is better established here than on the Atlantic slope of South America. Tools, machinery, and railway material are largely imported from the United States.

The sanitary influence of the United States is extending from Panama down the west coast, and is making residence safe, obviating delay by long quarantines, and thus introducing a necessary condition of trade. Much American capital has been invested in mining, and American influence is growing. As in almost all parts of the world, the trade of the United States is

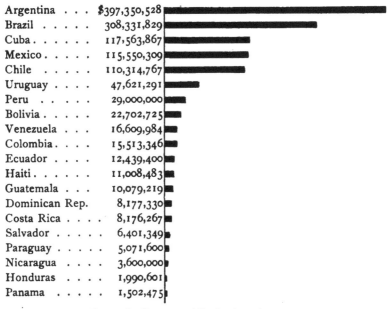

Argentina . . .	$397,350,528
Brazil	308,331,829
Cuba.	117,563,867
Mexico.	115,550,309
Chile	110,314,767
Uruguay	47,621,291
Peru	29,000,000
Bolivia	22,702,725
Venezuela . . .	16,609,984
Colombia. . . .	15,513,346
Ecuador	12,439,400
Haiti.	11,008,483
Guatemala . . .	10,079,219
Dominican Rep.	8,177,330
Costa Rica	8,176,267
Salvador	6,401,349
Paraguay	5,071,600
Nicaragua	3,600,000
Honduras	1,990,601
Panama	1,502,475

FIG. 248. Exports of Latin America, 1909

here also hindered by the predominance of foreign shipping. The opening of the Panama Canal will naturally lessen the traffic around the cape and increase the trade from the north. Railway development is greatest in Chile, the lines reaching out along the coast, parallel to the Andes, and, as already stated, across the range to Buenos Aires on the Atlantic coast. Valparaiso and Santiago, both in Chile, are the chief centers of the Pacific slope, while Callao is the port of Lima in Peru, and Guayaquil is the chief sea town of Ecuador. These countries are not of trifling size, and they have great variety in altitude, climate, and resources. Even Ecuador is nearly as large as the United

Kingdom ; while Chile is more than twice as large, and Peru is nearly four times as large, as this European country whose commercial importance is so vast.

340. Bolivia. This and Paraguay are the only countries of the continent which do not reach the sea. Bolivia consists in part of a lofty plateau and in part of broad slopes and plains

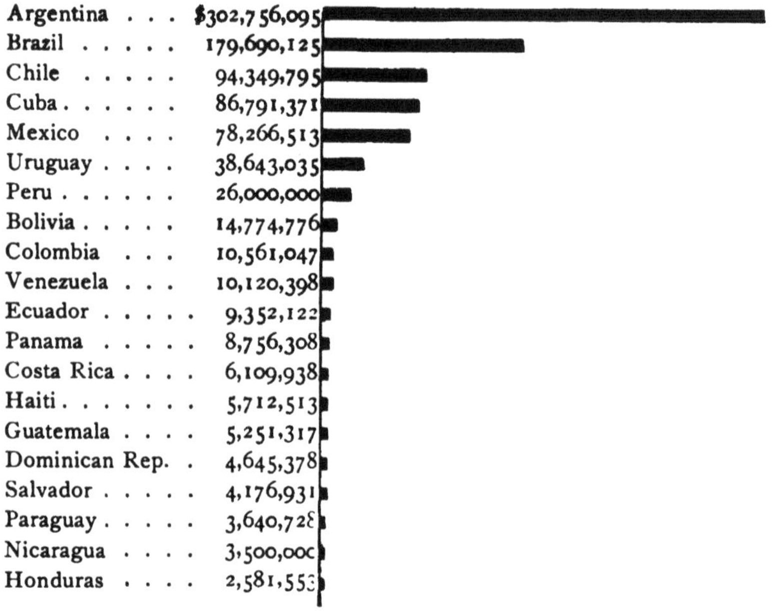

Argentina . . .	$302,756,095
Brazil	179,690,125
Chile	94,349,795
Cuba	86,791,371
Mexico	78,266,513
Uruguay	38,643,035
Peru	26,000,000
Bolivia	14,774,776
Colombia . . .	10,561,047
Venezuela . . .	10,120,398
Ecuador	9,352,122
Panama	8,756,308
Costa Rica	6,109,938
Haiti	5,712,513
Guatemala	5,251,317
Dominican Rep. .	4,645,378
Salvador	4,176,931
Paraguay	3,640,728
Nicaragua	3,500,000
Honduras	2,581,553

FIG. 249. Imports of Latin America, 1909

drained by the Madeira, the great southern branch of the Amazon. It has large wealth in copper, tin, and silver. The range of altitude introduces agriculture of both temperate and tropical types. Among routes for foreign trade, three are here named : (1) from Arica on the Peruvian coast to La Paz, the capital, a rail line being under construction ; (2) the Amazon route, entirely by river navigation, to the port of Pará in Brazil ; (3) to Buenos Aires by river, or by mule train to the Argentine frontier, and thence by rail.

341. Northern countries. These are Colombia, Venezuela, and British, Dutch, and French Guiana. Colombia lies chiefly, and the others entirely, north of the equator, amid typical tropical conditions. Colombia is also on the west coast, but the chief port is Barranquilla on the Caribbean Sea. Both Colombia and Venezuela are in somewhat close trade relations with the United States. Together they occupy more than 1,000,000 square miles; and this great area has a population of about 7,000,000, equal to that of London and its environs. With diversity of surface, mountain, valley and plain, and with great mineral and agricultural resources, these countries await development. They have not progressed in orderly government, in facilities of transportation, and in modern industry, as have Chile and the Argentine Republic. There has been, however, in recent years much improvement in these fundamental conditions of commerce. The total foreign trade of the two countries is somewhat over $50,-000,000, in which the United States has a larger share than in the trade of most South American countries. Venezuela alone has 2000 miles of coast line and many harbors. One port is on the Orinoco, 373 miles from the sea. Ocean vessels enter Lake Maracaibo by a broad strait, 34 miles long.

Venezuela has been in a disturbed state, but is now fulfilling her financial obligations to various European powers,. and is developing in harbor facilities, in railways, in telegraphs, and in education. Several steamship lines maintain regular service between Venezuelan ports and New York. All the usual tropical products are or may be raised, and the uplands and mountain slopes offer conditions suited to the cereals, roots, and fruits of the temperate zones. British, Dutch, and French Guiana, with an aggregate territory considerably greater than the United Kingdom, is the only South American region under European rule, contains but few people, and is but slightly developed. This, like other luxuriant tropical lands, will be capable of vast production when subdued and utilized by progressive people.

342. West Indies. Greatest and richest of these is Cuba, now an independent republic in somewhat close relations with the United States. The foreign trade amounts to about $200,000,000, of which two thirds is with the United States.

Nearly $150,000,000 of American capital are here invested, mainly in railways, sugar and tobacco plantations, real estate, and banking. Sugar and tobacco are the chief exports. The Bethlehem Steel Company of the United States now has the ownership of deposits of iron ore near Santiago, which are estimated to contain 75,000,000 tons.

Haiti occupies the western part of the island of that name, and trades in coffee, cocoa, and logwood, chiefly with the United States. There are cables to Cuba and South America, and steamships running regularly to New York. The Dominican Republic is in the eastern half of the island. The foreign trade amounts to $15,000,000, and about half is with the United States. Products and resources are in cocoa, tobacco, coffee, bananas, and in forests of cabinet woods and of yellow pine. Irrigation, mining, highways, education, and sanitation are undertaken, and the sloth and turbulence of former days are giving way to peaceful arts and productive industry.[1]

The Danish, Dutch, and French West Indies are small, have small trade, and illustrate no new principles of commerce. Jamaica, one of the greater islands, is a British colony of long standing, showing in a recent year imports of $25,000,000, due in part to rebuilding Kingston, the capital, following disaster. Fruit exports, mainly bananas, amount to $6,000,000, nearly all taken by the United States.

The Bahamas are perhaps the only British colony having no direct connection with the home country. The Bermudas do not belong to the West Indies, but may be mentioned here as a British possession whose trade with the United States is several times greater than with the United Kingdom. The most northerly of coral islands, they have a warm climate, furnish vegetables and bulbs for the markets of New York, and share in an important way in tourist traffic. The cable communication is with Halifax.

343. Panama and Central America. Panama in 1903 became independent of Colombia and ceded to the United States a

[1] For Porto Rico, see sect. 226.

strip across the Isthmus known as the Canal Zone. The area is 32,580 square miles, about two thirds the size of the state of New York. The foreign trade is about $10,000,000, not including importations for canal construction. Bananas — about 4,000,000 bunches — are the largest export. The railway, 47 miles long from Colon to Panama, is used for canal construction, also for commerce with Andean countries and Central America.

FIG. 250. Drying coffee, Costa Rica

Much merchandise passes between New York and San Francisco. The Isthmus offers a link in cable communication between Buenos Aires, Valparaiso, and the west coast in the south, and the United States and Europe in the north.

Central America includes the five small republics, Guatemala, Honduras, Salvador, Nicaragua, and Costa Rica, and also British Honduras. All the republics except Salvador border both the Caribbean Sea and the Pacific Ocean. Passage by rail from sea to sea will soon be possible in these four countries, and in time the Pan-American Railway will join them all along

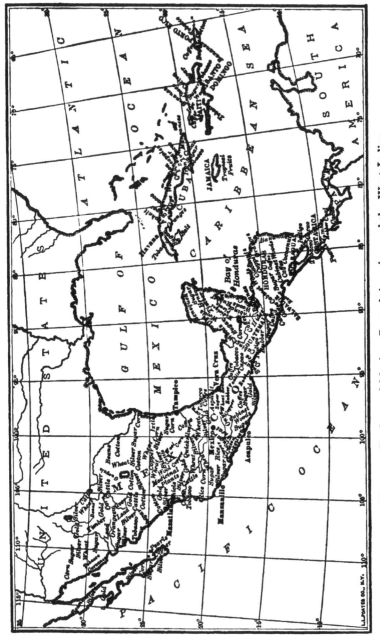

FIG. 251. Products of Mexico, Central America, and the West Indies

the great trunk line through North and South America. Social and political conditions and the natural resources are similar to those of Colombia and Venezuela. Good water, sewage, and the clearing of the jungle would make these lands as safe as the Canal Zone now is. There is a typical group of products : coffee, bananas, rubber, cocoa, dyewoods, cabinet woods, rice, sugar, indigo, and tobacco. Of the total foreign trade of the five republics ($56,000,000), the United States has about half. The largest center is Guatemala City, with about 100,000 people.

344. Mexico. Of the Latin-American republics Mexico is third in area and in volume of foreign trade and second in population. Having freed itself from Spanish rule nearly one hundred years ago, and having passed through various troubled conditions, the republic has, notwithstanding some vicissitudes, shown what a tropical land can do under a settled government. Laws have been so shaped as to make property safe, and foreign capital has flowed in to promote the various enterprises. Americans alone are said to have placed nearly or quite $800,000,000 in Mexico.

Mining interests are large, for the mineral resources are extensive. Thus Alaska and the Yukon, British Columbia, the Cordilleras of the United States and Mexico, and the Andes, all parts of the Pacific mountains of the American continents, are mineral territory. Mexico is one of the leading silver countries, the product of a recent year being $42,000,000. Gold and copper occupy the second and third rank, and the output of coal, lead, and petroleum is important. Foreign capital to the extent of $350,000,000 is invested in Mexican mines.

Here, as in the tropical Andean countries, the highlands offer conditions for products of temperate climates, while the lowlands are throughly tropical. The chief agricultural exports are, however, of tropical kinds : cocoa, coffee, sugar, vanilla, and henequen, the latter a fiber product. Crops of corn and wheat are considerable. The forests include pine, oak, tropical cabinet woods, and dyewoods. The foreign trade for the fiscal year 1907–1908 was $232,000,000, much the largest exchange being with the United States.

The government has undertaken irrigation works, and many water powers are being developed for manufacturing. A single reservoir in the Federal District of Mexico gathers the waters of a dozen streams and will provide for 236,000 horse power.

The railway system is growing rapidly and now includes about 15,000 miles. The greater part of the railways is under the virtual control of the government. Systematic development is planned so as to join the interior areas of production with the

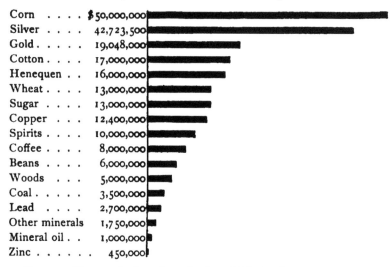

Corn	$50,000,000
Silver	42,723,500
Gold	19,048,000
Cotton	17,000,000
Henequen	16,000,000
Wheat	13,000,000
Sugar	13,000,000
Copper	12,400,000
Spirits	10,000,000
Coffee	8,000,000
Beans	6,000,000
Woods	5,000,000
Coal	3,500,000
Lead	2,700,000
Other minerals	1,750,000
Mineral oil	1,000,000
Zinc	450,000

FIG. 252. Plant and mineral production of Mexico in an average year

various seaports. The Tehuantepec railway is important for traffic in merchandise from the United States. This road was opened in 1907, and in the following year its traffic rose to $38,000,000.

Mexico has 1600 miles of coast line on the Gulf of Mexico and 2800 miles on the Pacific Ocean. Fifty-five places are counted as ports. Of these Vera Cruz is the chief, Tampico being the other leading Gulf port. On the Pacific, Mazatlan, Manzanillo, and Acapulco may be named. Manzanillo is as yet a small town, although ancient as a port. Important harbor works are under construction, and the breakwater is said to be one of the largest on any shore, exceeding those at Plymouth,

England, and Cherbourg, France. Posts and telegraphs are well advanced, and several wireless stations have been established. Mexico well merits the name which has been given to all Latin America, — the land of opportunity.

NOTE. The following, reproduced in abbreviated form from a publication of the Pan-American Union, states some of the measures that would aid in extending the trade of the United States in Latin-American countries :

1. The appointment of men of culture, tact, and energy as ministers and consuls.
2. The sending of business representatives who are gentlemen, and who speak Spanish, Portuguese, or French.
3. The manufacture of articles to suit the local Latin-American demand.
4. The giving of credit to reliable purchasers, as is done by European shippers.
5. The use of greater care in packing goods for the long distance to be traveled and for the severe changes of climate.
6. The opening of North American banks in the principal cities of South America.
7. The inducing of young Latin-Americans to come to our technical and professional schools instead of going to those of Europe.
8. The popularizing, in our schools and colleges, of the study of the Latin languages, history, and institutions.
9. The early building of Pan-American railway connections.
10. The investment of North American capital in South America.

CHAPTER XXVIII

345. Oriental period. At the dawn of history the more advanced peoples were in Egypt, Mesopotamia, Persia, India, and China. Population and industry were gathered along the Nile, the Euphrates and Tigris, the Ganges and Indus, the Yangtse and Hoangho. Because the life of man was found so largely on rivers, the civilizations have sometimes been called *fluvial*, or *potamic*, from the Latin and Greek words for river. Primitive commerce was carried on within and among these countries by land routes, and the pack animal and caravan were the means of transportation. Paths led then, as they lead to-day, from China through Chinese Turkestan, by Kashgar and Yarkand, to Persia, Mesopotamia, and the Mediterranean. Another route extended from the plains of India over the high passes to Persia and the west. Exchanges were carried on between Mesopotamian countries and Egypt; and as the Syrian coast of the Mediterranean developed, it also came into caravan communication with the Euphrates and Tigris.

346. Mediterranean period. This period also has its dim traditional beginnings and thus runs parallel to the oriental trade already described. By 1000 B.C. the greater commerce of the world no doubt was to be found along the shores and on the waters of the Mediterranean. The Mediterranean period covers at least 2500 years, — from 1000 B.C. to 1500 A.D. It includes the trade of the Phœnicians, the Greeks and their colonies, the Roman people and Carthage, and closes with the enterprises of Pisa, Genoa, and Venice.

As people increased in numbers, and civilization grew in Europe, the silks, spices, and gems of the Far East were sought, and the traders of the Mediterranean, the Red Sea, the Persian

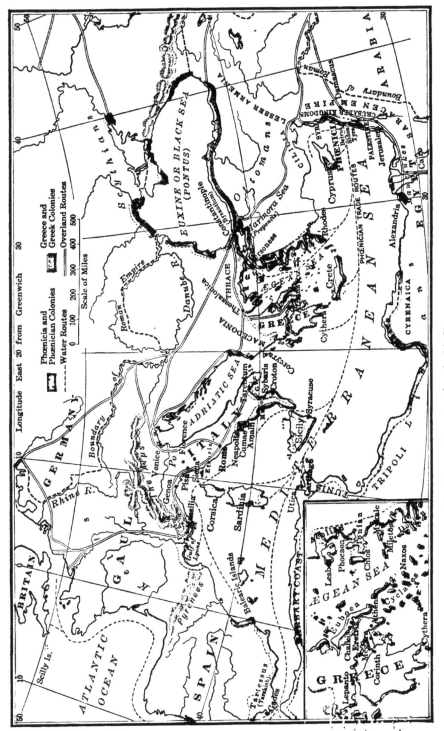

Regions and routes of early Mediterranean commerce

Gulf, and of the connecting land routes, found profit in making exchanges between continents. Thus the international barter of Asia passed into the intercontinental commerce of Asia and Europe, ships took largely the place of animals, the East and West were joined, and the world's trade was expanding. The fringe of trade was on the shores of the Pacific and Indian oceans in the east and touched the Atlantic border on the west.

Phœnicia occupied a narrow coast strip at the western base of Lebanon, 4000 square miles in area. It had ample harbors for small vessels and drew its shipbuilding supplies from the forests of the mountains. Its waters abounded in certain " shell-fish," or mollusks, which afforded the finest purple dyes. Phœnicia lay between the east and west, with Mesopotamia behind, and looking out upon Egypt, Greece, Italy, and Spain. A caravan route led from Mesopotamia by way of Damascus to Tyre, "a merchant of the people for many isles " (Ezekiel xxvii, 3). As early as the twelfth century B.C., according to Rawlinson,[1] Phœnicia held most of the world's carrying trade. Following the same author, their earliest trade by water was with Egypt, Cilicia, and Cyprus. They bought the merchandise of Egypt and sold it in Asia Minor, Greece, and Italy. Apparently Egyptian traders in their turn went far into the interior of Africa, for the traffic included ivory, ostrich feathers, gums, papyrus, fabrics, and glass. In Cyprus were found gems, and mines of copper, silver, and gold. Traders sailed to Rhodes and the islands of the Ægean, and from 1100 to 800 B.C. the coast of North Africa was occupied. Here arose Utica and Carthage. Some points in Sicily were settled, notably about the harbor of what is now Palermo. Other colonies were planted in Spain, as Gades (the present Cadiz), outside of Gibraltar. Mariners sped past Biscayan waters and found the Scilly Islands, and are believed to have worked and smelted the tin and lead of Corn-wall, selling these metals in Greece and in Asia. The men and women of Tyre and Sidon were skilled workers in textiles, metal-lurgy, and glass; and the carpenters, masons, and decorators, who

[1] Phœnicia, in Story of the Nations Series.

FIG. 253. Modern shipping in the harbor of Venice

aided Solomon in building the temple at Jerusalem, point to the same industrial ability. Thus the fundamental things in commerce find example in this ancient people. The picture is more complete if it be remembered that the agricultural Hebrews sent to Tyre, in return, wheat, barley, wine, and oil.

The Greeks also became active colonizers and traders, and planted their settlements at many points in southern Italy and Sicily, where splendid temples and other ruins still bear witness to their presence and their enterprise. From the Greek cities of Sicily oil and wine were sold in Africa, and breadstuffs were sent back to the parent land. About 600 B.C. Greek navigators went into the western Mediterranean and soon founded Massilia, the present Marseilles. This was an act of peculiar commercial importance, because the place was near the mouth of the Rhone and opened the historic route northward, up the Rhone and Saône, and across to the Seine and the English Channel. The Greeks thus traded with the merchants of Gaul, buying tin, which had been brought from Britain, and carrying it to Mediterranean markets.

The Romans were not typically a trading people, but the supply of Rome itself called for much commerce, as is seen in the " corn " ships that brought the produce of the fields of the Nile to the imperial city. And in her great efficiency as a road builder, Rome made her contribution to the commercial operations of the world.

It remained for certain other Italian cities in medieval and early modern days to establish trade and to bring the Mediterranean period to its culmination. Among these were Amalfi, now a relic of the past, visited by the tourist; Pisa, now separated from the sea by the growth of the delta of the Arno, and superseded by Leghorn; Genoa, powerful centuries ago, eclipsed by its Adriatic rival, but now risen to be first of Italian ports; and Venice, whose " glory and decline " form one of the great chapters in human history.

Venice, founded in the fifth century A.D., began about 1000 A.D. to trade with the Mohammedans. She took part in the Crusades and extended her power in the East, establishing her ports not only on the eastern Mediterranean, but in the Black Sea regions. Her power was greatest in the fifteenth century, when she had more than 3000 ships and was the sea power of the world (see map opposite). Like Phœnicia, but on a larger scale, Venice interchanged the merchandise of Asia and Europe; and her decline began and moved swiftly, when the Portuguese navigator, Vasco da Gama, found a way to India around the Cape of Good Hope. Venice then found herself aside from the new trunk route of the world's trade.

347. Atlantic period. The finding of an all-water route to India, and the growth of power and enterprise in Spain, Portugal, Great Britain, France, and Holland, transferred the seats of commerce to the neighborhood of the English Channel and the North and Baltic seas. As early as the thirteenth century, however, several towns of North Germany founded what has become known as the Hanseatic League, for the protection and fostering of commerce. Among these towns were Lubeck, Cologne, and Hamburg, and the League at one time included nearly 100 towns on or near the North and Baltic seas, mastering these waters and having important " factories " or trading establishments in Bruges, London, Bergen, and Novgorod. They built roads, fostered civilization, vied with great nations in power, and came to their decline in the sixteenth century.

The great discoveries of this period soon resulted in many colonies in the New World, and the British, Dutch, Spanish, and Portuguese became the commercial nations. As transatlantic settlements grew, the coasting trade became of less relative importance, and a truly oceanic trade was ushered in. Europe was long to remain the primary center, as indeed it is to-day, in the volume of its commerce. But the western world, under the leadership of the United States, has largely equalized manufactures and trade as between the Old World and the New.

The construction of the Suez Canal has given new importance to Mediterranean trade, and the revival is especially felt in such ports as Genoa and Marseilles. This revival is due in

the
nt
the
but
Asia
hen
lili
seli

to
rtu-
ars
ind
in,
be-
os-
ie,
oo
se
h-
lit
r,

ly
d

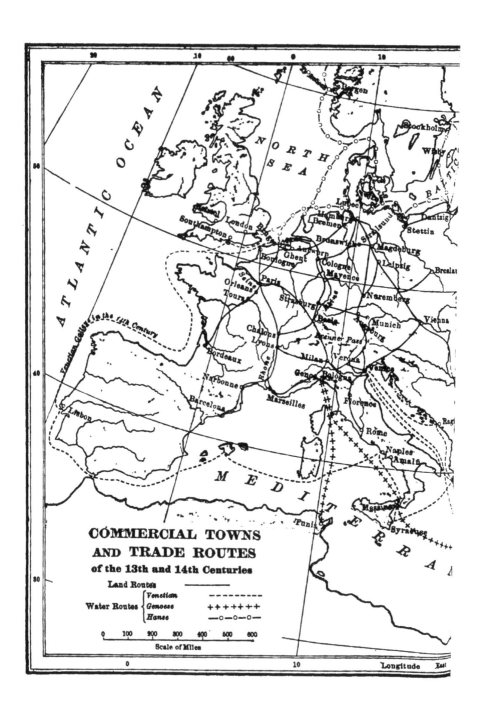

COMMERCIAL TOWNS
AND TRADE ROUTES
of the 13th and 14th Centuries

Land Routes

Water Routes ⎰ Venetian   – – – – – – –
            ⎱ Genoese   + + + + + + +
            ⎰ Hanse   —o—o—o—

0   100   200   300   400   500   600
Scale of Miles

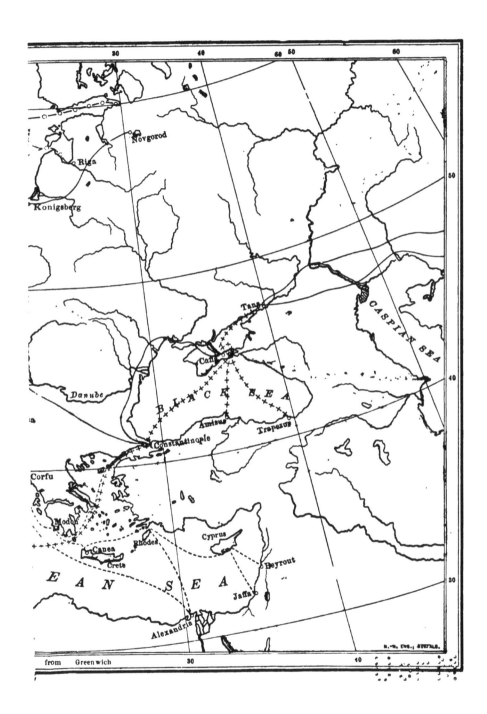

30   40   60 50   60

50

Novgorod

Riga

Konigsberg

CASPIAN SEA

Tana

40

Caffa

Danube

B L A C K   S E A

Amisus

Trapezus

Constantinople

Corfu

Modon

Cyprus

Rhodes

Canea

Beyrout

Crete

E A N   S E A

Jaffa

30

B.-N. ENG., BUFFALO.

Alexandria

from   Greenwich   30   10

part to the canal, but arises also from the new industrial progress of France, Italy, Austria, Egypt, and the Levant. The ascendancy, however, will not pass from the Atlantic to the old home of commerce. Should the Atlantic lose the primacy, it will move not eastward, but westward.

348. Pacific period. Every part of the Pacific shore line is now seeing a new development. About 1850, soon after the United States gained a firm foothold on the western ocean, growth began, and is now rapid, not only in the three coast states but in Alaska. Since the Canadian Pacific Railway reached the Pacific, development has gone on by leaps and bounds on the western border of Canada. Ten Latin-American republics with their new life border the Pacific. On the south and west the story is the same, as shown by New Zealand, Australia, the East Indian islands, and Japan; and on the continental border of Asia, where China is struggling up into industrial and commercial strength, and where Russia, with various vicissitudes, has established herself.

Into this circuit of nations the Isthmian Canal will open a way for the ships of Europe and eastern America. Such lands as the western United States, Canada, all the Latin-American republics, and Australia will increase in population, while western Europe is approaching its limit. Should China arise to commercial activity, it would mean the entrance into trade of a people as numerous as the entire population of Europe, supported by equal or greater natural resources. It may, however, be considered as unsafe to predict at this stage whether the Pacific Ocean will take the place of the Atlantic as the theater of the world's greatest trade.

349. Types of commercial expansion. We observe (1) the expansion of manufacture and trade among the progressive nations of Europe, the United Kingdom, Germany, Belgium, France, Switzerland, and Austria-Hungary; (2) the revival of trade and the entrance of modern ideas into lands of former commercial greatness, such as Italy, Spain, Greece, and Egypt; (3) the development of vast new countries of small aboriginal populations : such are the United States, Canada, Argentina, Australia, and New Zealand ; (4) the opening of populous and backward old lands, of which India, China, and Japan are examples ; (5) the conquest of the tropics, offering new problems, such as the safe living of white men in hot regions and the training of the natives to work and trade. There will always be great differences between tropical and temperate products, and hence exchanges will be large and trade will be enduring. (6) Finally, may be noted national ambition and the desire for exploration, seeking to know and use all regions, as is shown in the partition of Africa. Eager science also finds new materials and new uses of old things, and brings them within the grasp of all.

350. Social and moral effects of commerce. The reaction of industry and trade modifies the habits, judgments, and policies of men and nations. Abundant examples are at hand, but the influence is so powerful and so general that coming years will make the results far more evident than they now are. Hints of such changes will now be given.

1. Order and security. Several years after the war with Spain disorder arose in Cuba, and property was destroyed by lawless hands. The holders of American and European investments in that island at once strongly pressed the United States to intervene and protect property interests. This was done, and American soldiers and a provisional governor took possession until order was restored. According to Chisholm the wealth derived from the tin mines of the Malay Peninsula " has been the chief means of converting a proud and lawless people into a submissive and orderly community." The Isthmus is an important example in the New World.

2. Financial integrity. Venezuela is an undeveloped country, with an unstable people. For years they tolerated a grasping and dishonest dictator as the head of their government. Just debts owed to European nations were withheld or repudiated. The pressure brought by these nations has given new security, the corrupt ruler was returned to private life, and the turbulent

public has learned, in part at least, the lesson of common honesty. Nations must keep their credit good in order to negotiate needed loans in the money markets of the world.

3. Education. Modern manufacture, transportation, and trade involve the use of technical appliances, which, being applications of modern science, require an advanced degree of education. The chemist, the electrician, the mining geologist, the civil engineer, the economist, the consular representative, and the agricultural expert must lay their foundations of knowledge in the schools. The rapid rise of Germany is attributed in part to the number and thoroughness of the technical schools and to the advanced condition of the German universities.

4. Economy. This is found in the division of labor which transportation makes possible ; in large aggregations of capital and labor under one capable management ; in the best and most exact and scientific use of resources, including the use of waste materials for by-products ; in equalizing charges and making a common world market for products ; and in intelligent use of materials and merchandise to meet those higher standards of living which cultivated people now find necessary to their comfort and success.

5. Development of sympathy. This arises in part through modern communication, which makes the world instantly acquainted with conditions of famine, disaster, or persecution. When the news came of disaster in San Francisco and Messina, the sympathies of the world were given instant opportunity for action. Oppression in Armenia, or cruelty to natives in the Kongo, arouses the feeling and elicits the protest of the world, and thus develops the common feeling of the human race in a degree unknown before the days of modern commerce.

6. Unified public sentiment. This is akin to development of sympathy, and, like it, is possible through instantaneous communication. The world sits in judgment, over every morning's paper, upon what men and nations do. Thus unity arises which becomes crystallized in international arrangements. Postal unions, the Hague tribunal, temporary courts of arbitration, international

law, submarine cables, and swift ships are the instruments of unity
for the world. Isolation has been called the mother of barbarism,
while communication and trade bring nations and men together,
often put evil to shame, and, by the light of publicity, establish
better things and promote the higher life of man.

7. Prevention of war. Industry requires peace and freedom
from the hazards of war. Strife among nations is wasteful of
wealth, and a modern war seems likely to become too costly to
be tolerated. The loss is in time, in valuable lives, in money
spent, in productive industry interrupted, in property destroyed,
in debts assumed, and in burdens laid on posterity. Thus economic motives tend to banish war, and the growth of neighborly
intercourse establishes peaceful ways of settling differences. It
is also possible that modern technical science will make war
so much more destructive of life than it now is, as to banish
it altogether from the experience of nations.

8. The problem of the future. Commerce is not altogether
peaceful or benevolent, and it must be admitted that modern
invention has put vast power into the hands of able and selfish
promoters of business. Thus advantages and difficulties come
together, as must be expected. But if the telegraph and railway
give advantage to the monopolist, they also make possible rational
protection of interests by the laborer and artisan, and they lend
equal aid to the citizen seeking to establish just regulation and
government. It remains for the future to show whether international trade warfare will give way to coöperation and adjustment,
and whether, in the end, each man and each country may be permitted to do what each can do best, for the greater good of all.

INDEX

Aberdeen, 310, 316
Abrasives, 173
Abyssinia, 414
Acapulco, 438
Adelaide, 409
Aden, 388
Adirondacks, 114; iron ores in, 61
Adjustment, illustrations of geographic, 215
Africa, 412–421; agriculture, 418, 419; climate, 412; coal, 88; countries by latitude belts, 414–418; European dependencies, 413; mineral resources, 419, 420; partition, 414; people, 412; physical features, 412; position, 412; productions and trade, 412, 413; transportation and centers, 420, 421
Agricultural explorers, 3
Agriculture, Africa, 418, 419; Argentina, 428; Australia, 408; Canada, 292–294; distribution of products, United States, 121; England, 305, 306; experiment stations, 266 (map), 267; France, 322, 323; Germany, 337, 338; Hungary, 374, 375; implements used in, 214, 215; India, 391; Italy, 359, 361; Mexico, 437; Netherlands, 347; population engaged in, United States, 103; Russia, 379, 381; Scotland, 314, 315; United States (see wheat, cotton, plant products)
Alabama, coal, 82, 83; iron ores, 61
Alaska, gold, 175; grains, 267; national forests, 143; salmon, 151; seal fishing, 156; telegraph service, 254; trade, 276; wheat, 4
Albert Nyanza, 418
Alberta, wheat in, 293
Aleppo, 387, 388
Alexandretta, 386
Alexandria, 415, 416, 420
Alfalfa, for cattle, 43; for swine, 147
Algeria, 332, 413, 414, 420, 421
Algiers, 421; wheat, 5, 19

Allegheny uplands, in Pennsylvania, 114
Almeria, 365
Alps, 321; Austrian, 371; routes from Italy, 358
Amalfi, 443; drying macaroni at, 358
Amazon River, 427
Amsterdam, Holland, 348; New York, knit goods at, 207, 208
Andalusia, 364
Angola, 413
Animal industries, United States, 145–159; bees and products, 154, 155; feathers, 156; fisheries, 150–153; furnishing motive power, 157; furs, 156; hogs, 146, 147; leather, 155, 156; organization, 157, 158; poultry and eggs, 148; raw materials, 97; sheep, 145, 146; silk, 157; various, 159; wild game, 148–150; wool, 155
Anthropogeography defined, 92
Antioch, 385, 387
Antwerp, 328, 334, 356; Canadian Pacific Railway piers, 301; view of harbor, 333
Appalachian region, coal fields, 82, 83; description of physical features, 113, 114; forests, 139, 143; petroleum, 161, 162; plateau, 114; relief map, 112; routes of trade, 224, 230; water power, 193
Apples, in United States, 127
Arabia, 385, 388; Arabia Felix, 388
Arbitration, courts of, 447
Archangel, as wheat port, 18
Ardennes, the, 321, 333
Argentina, 426–430; agriculture, 428; cattle, 48; exports, 428; imports, 429; physical features, 427; railways, 429; trade with United States, 279; wheat, 3, 19, 426
Arica, 432
Arizona, copper, 176, 177; gold, 175; silver, 176; subtropical fruits, 126
Arkansas, coal, 83; thermal waters, 201

Lightning Source UK Ltd.
Milton Keynes UK
UKHW010646201218
334320UK00013B/655/P